The Death of a Great Company

Also by W. Julian Parton:

The Story of the General Crushed Stone Company

THE DEATH
OF A GREAT COMPANY

REFLECTIONS

ON THE DECLINE AND FALL

OF THE

LEHIGH COAL AND NAVIGATION

COMPANY

W. Julian Parton

**Canal History and
Technology Press**

Copyright © 1986 by W. Julian Parton

Published by
Canal History and Technology Press
National Canal Museum
30 Centre Square, Easton PA 18042-7743

Second Edition, 1998

Library of Congress Cataloguing in Publication Data

Parton, W. Julian, 1917 –
        The death of a great company : reflections on the decline and fall of the Lehigh Coal and Navigation Company / by W. Julian Parton.
        p.   cm.
        Reprint, 1st ed. published: Easton, Pa. : Center for Canal History and Technology, 1986.
        Includes bibliographical references and index.
        ISBN 0–930973–18–6 (pbk. : alk. paper)
        1. Lehigh Coal and Navigation Company — History.   2. Coal Trade — Pennsylvania — History.   3. Coal — Pennsylvania — History.   I. Title.
        HD9549.L46P37   1998
        338.7'622334—dc21   98–17707
                                                                            CIP

**Canal History and
Technology Press**

CONTENTS

The Author vii

Acknowledgements ix

List of Illustrations and Tables xi

Introduction by Thomas Dublin xv

Prologue 1
    Founding Years
    "The Switchback"

The Lehigh Coal and Navigation Company 3
    Facilities, Location and Assets of the Lehigh Coal
        and Navigation Company in the Late 1930s
    The People of the Panther Valley
    "The Old Company" as a Corporate Citizen
    General Industry Conditions in the 1920s

The White Years (1938–1953) 33
    Pre-War Years (1938–1941)
    The War Years (1941–1945)
    The Post-War Years (1946–1953)
    Problems at the Mines
    Employee Problems
    The Disastrous Years (1949–1953)

The Dodson Years (1954–1959) 69
    Changes at the Lehigh Navigation Coal Company
    Parton's Plan to Operate the Mines
    Decision to Discontinue Lehigh Navigation Coal
        Company Mining Operations
    Panther Valley Coal Company Lease of Lansford Colliery
    Coaldale Mining Company Lease of Coaldale Colliery
    Merger of Panther Valley Coal Co. and Coaldale Mining Co.
    Expansion and Diversification under Dodson

The Thompson Years (1959–1969) 99
    The Plan of Reorganization
    The New Lehigh Coal and Navigation Company
    L.N.C. Corporation

Epilogue 107

Notes 113

Index 119

THE AUTHOR

W. Julian Parton has a vital interest in the history of the Lehigh Coal and Navigation Company. He was part of it. Fresh from the University of Washington graduate school, he started working for its subsidiary company, the Lehigh Navigation Coal Company, as a mining engineer. In the following fifteen years he held positions as supervisor of flotation projects, planning engineer, colliery superintendent, assistant general manager, vice president, and finally president.

His coal mining career seemed dashed to pieces when the directors of the Lehigh Coal and Navigation Company decided to go out of the business of mining anthracite in 1954 because of heavy operating losses and a militant labor force at the coal mining subsidiary.

After serving as administrator of the mining properties for six months, he and three associates organized their own company, The Panther Valley Coal Company, and leased the Lansford and Nesquehoning Collieries. The new company did well, initially, when the miners responded to the "work harder - increase productivity" goal. However, one of the other previously idled collieries was leased by the Coaldale Mining Company and the resulting competition forced the owners of the Panther Valley Coal Company to sell their interests.

Discouraged, the author decided to leave the coal industry in 1955 and entered another phase of the mining business. He was fortunate once again to become president and, later, chairman of the board of his new company, The General Crushed Stone Company. However, he carried fond memories of the company that had given him his start in business. He had gained valuable experience and knowledge from his employment with the Lehigh Navigation Company which served him well in his later career.

Thirty years after leaving the anthracite region, now retired, the author decided to put in print his recollections of the declining years of the Lehigh Coal and Navigation Company. "The Death of a Great Company" is his first attempt at a publication of this nature. His previous publications were articles pertaining to various technical aspects of mining and processing coal and crushed stone.

W. Julian Parton was born in Pottsville, Pennsylvania, on January 9, 1917. He attended the Pottsville High School and then the Pennsylvania State University where he received a B.S. degree in Mining Engineering in 1938. After graduation at Penn State, he attended the University of Washington as a research fellow and was awarded an M.S. degree in Mining Engineering in 1939. In 1944, The Pennsylvania State University awarded him an Engineer of Mines degree as a result of original research work he accomplished at the Lehigh Navigation Coal Company.

He and his wife, the former Nancy Joy, divide their time between their homes in Easton, Pennsylvania, and St. John, the U.S. Virgin Islands.

ACKNOWLEDGEMENTS

Many people were most helpful in providing the author with historical information and photographs to assist in piecing together the drama which took place when the Lehigh Coal and Navigation Company was fighting for survival. H. Louis Thompson, the last president of the company, contributed a great deal by sharing his recollections. His suggestions and criticisms were invaluable. T.R. Berger, public relations officer for the company during the crucial years, made available his former files on the company. He guided the author from the start of this undertaking, and was most generous with his time and skills in reviewing and editing the text. Attorney William B. Joachim kindly reviewed the text to determine the accuracy of the contents.

Clarence Dankel, the "last" coal company employee, made his collection of memorabilia available to the author. William Richards, the "Panther Valley Historian," likewise was generous with his collection of historical information, slides and photographs. John Gunser, curator of the Asa Packer Mansion, Jim Thorpe (formerly Mauch Chunk), assisted in reliving the early history of the company when this city was the center of activities because of the gravity railroad and the Lehigh Canal.

George Harvan, Lansford photographer, made available appropriate photographs taken of the company facilities and employees. Greg Buchala was also a contributor of information and photographs.

The Center for Canal History & Technology at Easton, Pennsylvania, opened its doors to the author and made available photographs, documents, and reports on the company. Steven Humphrey, museum director, and Lance E. Metz, program director-historian, were most supportive.

To all who helped in any way, the author wishes to express sincere appreciation. Thank you.

List of Illustrations

Founders of the Lehigh Coal and Navigation Company,　2
　Josiah White and Erskine Hazard.
Early view of Mauch Chunk, now Jim Thorpe,　3
　showing Lehigh Canal in foreground.
Original chutes loading canal boats at Mauch Chunk.　3
Lehigh Canal, showing Lehigh Coal and Navigation Company　4
　boat returning after carrying coal to market.
Hauto Tunnel, completed in 1872. It connected with the Central　6
　Railroad of New Jersey and eliminated the need for the gravity
　or "Switchback" railroad to transport coal to Mauch Chunk.
Schematic sketch of Lehigh Coal and Navigation Company　7
　gravity railroad.
Base of the Mt. Pisgah Plane of the Switchback Railroad,　8
　Mauch Chunk, Pennsylvania.
Down Track of the Switchback Railroad alongside Mauch　10
　Chunk Creek.
Side by side the canal and railroads operated close together　12
　until the canal was finally abandoned.
Lehigh Coal and Navigation Company coal measures　14
　and adjacent areas.
Tamaqua Colliery and froth flotation plant (right), Lehigh　16
　Navigation Coal Company. Built 1922; razed June 25, 1958.
Coaldale Colliery, Lehigh Navigation Coal Company.　16
　Built 1922, razed 1962.
Lansford Colliery, Lehigh Navigation Coal Company.　17
　Built 1925; burned 1960.
Lansford shops of Lehigh Navigation Coal Company　17
　employed 350 men at its peak. Initial construction, 1872.
Nesquehoning Breaker, Lehigh Navigation Coal Company.　18
　Built 1908; razed 1948.
Hauto storage yard enabled the Lehigh Navigation Coal Company　18
　to stock up to 360,000 tons of coal during periods of low demand.
Main office of the Lehigh Navigation Coal Company, Lansford.　19
　Built in 1872; enlarged in 1880s; destroyed by fire in December,
　1975. Headquarters for the "Old Company's Lehigh" mining
　operations.
Lehigh and New England Railroad steam locomotive　20
　hauling a train of coal cars across the Lehigh River at Palmerton.
Dedication ceremonies for the Summit Hill-Lansford-Coaldale　22
　Community Pool and Park.

Annual outing of the Old Timers Club 22
    (workers of Lehigh Navigation Coal Company with 50 years
    or more of service). President Evans serving the guests.
Happier times at Old Timers Club outing at Greenwood Lake. 23
    R.V. White, Evan Evans and J.B. Warriner "singing along."
Samuel D. Warriner, President of Lehigh Coal and Navigation 26
    Company 1912–1937; Manager 1912–1941.
Jessy B. Warriner, President of Lehigh Navigation Coal Company 28
    1930–1946; Manager 1942–1948.
The Big Three: W.H. Edwards, President, Lehigh & New 29
    England RR; R.V. White, President, Lehigh Coal & Navigation
    Co.; J.B. Warriner, President, Lehigh Navigation Coal Co.
Robert V. White, President of Lehigh Coal and Navigation 33
    Company 1938–1959.
Original Split Rock Lodge at Lake Harmony, Pennsylvania. 40
Froth Flotation Plant at Tamaqua Colliery, designed to recover 43
    fine sizes of anthracite and prevent stream pollution.
Group inspecting the 40-foot and Mammoth stripping excavation. 43
Evan Evans, President, Lehigh Navigation Coal Co. 1947–1952. 45
Mr. Robert Taney, supt. of the Lelite Plant, describing Lelite. 48
    The Old Company Lehigh was a pioneer in this development.
New diesel locomotive of the Lehigh & New England RR, 49
    part of the modernization program.
New method of mining, employing long hole drilling and 51
    blasting techniques.
Lehigh Navigation Coal Company sketch showing development 52
    in rock required to divide the "lift" distance between levels
    in thick heavy pitch mining.
Tamaqua truck scale. 53
Narrow gauge railroad from Nesquehoning Colliery to Lansford, 53
    completed in 1948. Railroad made it possible to close the
    Nesquehoning Breaker.
Miner and his "buddy" drilling in a breast at a mine of the 54
    Lehigh Navigation Coal Company.
Lehigh Navigation Coal Company cross section showing veins 56
    in Panther Creek Valley between Lansford and Tamaqua.
Lehigh Navigation Coal Company officers in June, 1953: 66
    Glenn O. Kidd, President; Evan Evans, Chairman of the Board;
    Richard Newbold, Vice President of Sales.
C. Millard Dodson, President, Lehigh Coal and Navigation 71
    Company, 1954–1959; Manager, 1953–1959.
Lehigh Navigation Coal Company President W. Julian Parton 73
    and Vice President Joseph Crane studying Coaldale Breaker.

International U.M.W.A. officers involved in Panther Valley   75
    labor problems in 1953: Thomas Kennedy, vice president;
    Santo Volpe, union official; John L. Lewis, president;
    Martin Brennan, president, District 7, U.M.W.A.
Mass meeting of Tamaqua Colliery miners in 1954.   79
Panther Valley officials James Fauzio, Joseph Crane,   82
    W. Julian Parton, Frank Fauzio.
Evan Thomas, Sales Manager, Lehigh Navigation Coal Company,   84
    inspecting the first car of coal from the Panther Valley Coal
    Company, October 6, 1954.
Happy Nesquehoning miners returning to the "pits" after   85
    reopening of the mines under lease to the Panther Valley
    Coal Company on October 6, 1954.
Miners handing in lamps to James Gover, lampman at   86
    Nesquehoning Mine at end of first day mines reopened.
H. Louis Thompson, President, Lehigh Coal and Navigation   99
    Company, 1960–1968; Manager, 1954–1970.
Fauzio Brothers' stripping operation and equipment.   109

Lehigh Coal and Navigation Company General Map   Inside back cover

TABLES

Table I: The Lehigh Coal & Navigation Company   15
    securities owned, December 31, 1939

Table II: The Lehigh Coal & Navigation Company and Subsidiary   36
    Companies Consolidated Balance Sheet, December 31, 1938.

Table III: Results of Lehigh Coal and Navigation Company and   46
    Subsidiaries in Post-War Yars (1946–1954).

Table IV: The Lehigh Coal & Navigation Company and Subsidiary   87
    Companies Consolidated Balance Sheets, December 31, 1954
    and 1953.

Table V: The Lehigh Coal & Navigation Company and Subsidiary   102
    Companies Consolidated Balance Sheets, May 31, 1962.

INTRODUCTION

The Death of a Great Company offers a rare microcosmic view of the broader process of deindustrialization that has transformed the American economic landscape in the post-World War II era. In the forty years between 1950 and 1990 the proportion of workers in the United States employed in manufacturing declined from 26 to 18 percent.[1] Employment in anthracite and bituminous coal has plummeted, followed only a few years later by similar declines in steel and auto. Increasingly the United States is importing its fuel, its steel, its autos, and its textiles. And the recently signed North American Free Trade Agreement (NAFTA) and the last round of talks on the General Agreement on Tariffs and Trades (GATT) will both reinforce the internationalization of production for the nation's consumer markets. Manufacturing jobs will undoubtedly continue to comprise a shrinking share of the American labor force in years to come.[2]

Thus the experience of industrial decline in the anthracite region of Pennsylvania in the postwar decades foreshadows parallel declines in other industries and speaks to broader patterns of economic and social change in the United States. Julian Parton was uniquely situated to participate in and observe this decline as it affected a single large firm, one of the nation's leading anthracite producers, the Lehigh Coal & Navigation Company. He was an "insider" in that he worked for fifteen years as an engineer and manager in the firm's mining operations, yet he was also something of an "outsider," as one who continued to advocate the production of coal in the face of a corporate takeover that transformed a coal and railroad firm into the speculative plaything of Wall Street investors. Writing thirty years after he left the company, Parton was able to distance himself from the swirl of events and report on the company's demise with an unusual detachment.

His independence came from his engineering training and his commitment to producing coal. He was raised in the anthracite region, graduating from high school in Pottsville. His father, however, had no direct connection with mining operations, working rather in accounting, sales, and management. With degrees in mining engineering from Penn State and the University of Washington, Parton had not initially intended to return to his native region. Failing to land a job in western metal mining, Parton returned to Pennsylvania. A recommendation from the head of the mining school at Penn State led to an interview at Lehigh Navigation Coal and Parton's first job. He clearly had a knack for the work and was soon recognized for more than his engineering talents. He was appointed superintendent of the company's Nesquehoning Colliery, then assistant general manager, and, in 1954, president of Lehigh Navigation Coal.[3]

As becomes clear in Parton's account, he found himself caught between unionized mineworkers, determined to defend their interests, and a Board of Managers more concerned with the profits to be made by selling off company assets than in mining coal. His first action as president was to suspend production indefinitely, while he worked out with the United Mine Workers of America (UMWA) a plan to reopen the mines on a profitable basis. With demand for anthracite declining steadily, there was no way that all of the company's mines, stripping operations, and breakers could remain in operation.

Parton came up with a plan, but in the end he could not overcome five decades of mistrust and labor strife in the Panther Valley.[4] He thought he had agreement on a plan that would maximize employment by closing down all stripping operations and opening all underground operations instead. However, the production anticipated would have only been great enough to keep one breaker in operation, that at Coaldale. Under the plan, coal mined in Nesquehoning, Lansford, and Tamaqua would be transported by locomotives to Coaldale for processing. While the plan received the support of UMWA leadership and the General Mine Committee, some rank-and-file miners had reservations. In final voting, five locals voted to return to work, one opposed the proposal, and another voted to return to work but opposed the new workplan offered by the company. In the end, when work resumed, members of the dissident local set up picket lines and miners refused to cross them. After two and a half weeks of union meetings, charges, and counter-charges, the company's board of managers met in Philadelphia and voted to discontinue mining operations in the Valley. The temporary suspension had become permanent.

Parton's account lays the onus for the final closing of the mines on intransigent mineworkers. And there is no doubt that in the course of events in May and June 1954 the tactics of dissident miners succeeded in keeping the majority of workers who had voted to return to work from doing so. Still, one wonders how long any resumption of mining would have lasted, given the priorities of the company's new stockholders and the sentiment of members of the Board of Managers. For, as Parton tells the story, financial developments played a determining role in the firm's demise. A group of Wall Street investors had singled out the Lehigh Coal & Navigation Company as a target for a takeover effort. They found that the book value of the company's assets far exceeded the total value of its stocks and were purchasing company stocks accordingly. As their stock ownership increased, they demanded representation on the company's Board of Managers. They were more interested in selling off the company's railroad assets and pocketing the proceeds than in continuing to produce and ship anthracite coal. Although the Board of Managers ordered Parton to develop a plan under which to reopen the mines, it is not clear that the company's new owners had much interest on mining coal. Parton's narrative indicates that the railroads, more than coal lands and operations, were the prize the new investors were seeking. Once they had been sold off, it was unlikely that the mining operations would prove particularly profitable given the ongoing decline in demand for anthracite. All in all, the labor troubles provided the occasion for closing the mines; it is very unlikely that mining operations would have continued for long even with the most accommodating of union miners.

Although the focus of Parton's account is on successive management teams at Lehigh Coal & Navigation and on the immediate crisis that led to the mine closings, he also offers readers rich evidence on the labor history of the Panther Valley. He discusses, for instance, the combative relations between workers and management that had evolved since the 1930s, demonstrating how developments gave workers considerable authority in the work process. Supervision at the coal face was minimal and workers came to view traditional practices as vested rights that they guarded jealously. In Parton's description of labor-management relations over two decades, we see solid evidence of the kind of "workers' control" that labor historian David Montgomery has ascribed to skilled labor in American industry in the early twentieth century.[5] Contract miners working at the coal face were paid on a piecework basis and were free to determine the best way to proceed with their work and to come and go as they liked. They commonly worked five or six hours, blasted from the face all the coal they needed to make up a day's "stint," and quit work, leaving the loaders to fill the mine cars from their chute. The miners—the élite of the mines' industrial workforce—developed a sense of independence that made them hard to control and set the stage for conflict that Parton describes at the Lehigh Navigation Coal Company in 1954.

This independence and quickness to take offense on the part of company miners drew upon what Parton terms a "deep-seated feeling of hostility against the Company." (p. 55) As Parton notes, workers at LNC in the 1950s were commonly the sons and even grandsons of former workers. The attitudes toward the company developed in the great strikes of 1902, 1922 and 1925, and honed still further in the equalization movement of the 1930s, were passed down across the generations and made rank-and-file miners skeptical of the company's justifications for its actions.

Parton is a keen observer and judge of circumstances and his account of the conflict that led to the company's closing has much of value to students of history. He does not draw extensively on contemporary records and he clearly played a particular managerial role throughout the period of crisis; still his account has real value for an understanding of industrial decline in the Panther Valley in particular and in the nation as a whole. His evidence of the interplay of market forces, issues at the point of production, and financial pressures emanating from Wall Street demonstrated that deindustrialization is the product of complex economic forces and not simply the result of trade union intransigence in the face of changing market conditions. Deindustrialization emerges within a concrete institutional setting and reflects the priorities and choices of stockholders and corporate managers within an ever-changing national and world economy.

The problems described by Julian Parton are very much with us in the mid-1990s, some three decades after the final liquidation of Lehigh Coal & Navigation. *The Death of a Great Company* continues to address significant questions that the American economy faces today and will continue to face in the future. In the declining former coal towns of the Panther Valley—in Tamaqua, Coaldale, Lansford, Summit Hill, and Nesquehoning—we see aging residents surviving principally on Social Security and black-lung compensation payments. We see as well decaying physical remnants of a once "Great Company," fading into a landscape scarred by a century and a half of underground mining. The Lehigh Coal & Navigation Company dominated the Panther Valley for some 140 years and its demise left the Valley with little economic basis for the future. Its departure leaves unanswered questions and perhaps insoluble problems. How to reconcile the imperatives of capitalist industrialization with broader human and environmental needs is the challenge still facing the nation as the United States prepares to enter the twenty-first century.

Thomas Dublin, 1995
State University of New York at Binghamton

NOTES

1. U.S. Department of Commerce, Statistical Abstract of the United States (Washington, D.C.: G.P.O., 1994), 412; U.S. Department of Commerce, Business Statistics, 1971 (Washington, D.C.: G.P.O., 1971), 67, 69.

2. Useful readings exploring the broader phenomenon of deindustrialization include: Paul D. Staudohar and Holly E. Brown, eds., Deindustrialization and Plant Closure (Lexington, Mass.: Lexington Books, 1987); John Gaventa, Barbara Ellen Smith, and Alex Willingham eds., Communities in Economic Crisis: Appalachia and the South (Philadelphia: Temple University Press, 1990); John T. Cumbler, A Social History of Economic Decline: Business, Politics, and Work in Trenton (New Brunswick, N.J.: Rutgers University Press, 1989).

3. Much of the biographical information comes from The Death of a Great Company, but it is supplemented by an interview with W. Julian Parton in his home in Easton, Pennsylvania, July 31, 1993. Transcript of that interview in possession of the author.

4. For a discussion of the form that labor strife took in the 1930s in the Panther Valley, see Thomas Dublin, "The Equalization of Work: An Alternative Vision of Industrial Capitalism in the Anthracite Region of Pennsylvania in the 1930s," Canal History and Technology Proceedings vol. XIII (1994):81-98.

5. David Montgomery, Workers' Control in America: Studies in the History of Work, Technology, and Labor Struggles (Cambridge, England: Cambridge University Press, 1979), 9–31.

PROLOGUE

One of the oldest companies in the United States, the Lehigh Coal & Navigation Company, formerly known in the coal industry for many years as "The Old Company," has ceased to exist. The people who founded it and those who followed were in many ways significant contributors to the development of America, particularly in the mining and transportation fields.

The story of the "Old Company's Lehigh" should be recorded before all the people who were involved in its better days pass from the scene. This is the goal of the author. Emphasis, however, is placed on the final active years of the Company when one thing after another went awry, leading to the liquidation of virtually all its assets, and wiping out 150 years of growth.

The liquidation of this great Company should be mourned as the passing of one who has lived a full life and whose efforts were productive and rewarding to family, friends, and associates. However, a company is not a mortal human being; it need not die measured against the fact that incorporation under the usual laws gives it a right to perpetual existence. Permanence and stability are characteristic features of a corporation. Why, then, was it necessary for the Old Company to be put to an untimely rest?

Founding Years

The accidental discovery of anthracite coal occurred in 1791 when a hunter named Philip Ginter[1] ventured to the summit of Sharp Mountain (now the Borough of Summit Hill) nine miles west of Mauch Chunk (now Jim Thorpe), in Carbon County, Pennsylvania. Ginter took the sample of "stone-coal" to Colonel Jacob Weiss, a former Revolutionary War purchasing agent, who was residing at Fort Allen (later renamed Weissport for him).

Colonel Weiss, Charles Cist, a well-known printer, Michael Hillegas, the first Treasurer of the United States, and others formed what was to be The Lehigh Coal Mine Company.[2] They took up 8 or 10 thousand acres of what was then undeveloped land, including the Sharp Mountain. Early attempts to mine and market the coal were not very successful for want of knowledge at the Philadelphia market as to how it should be used.

Josiah White and Erskine Hazard were manufacturing nails at the "Falls of the Schuylkill" in Philadelphia.[3] Interested in Lehigh coal, they purchased some of the first shipments which were transported by floating arks down the Lehigh and Delaware Rivers to Philadelphia. Their experimentations with the use of this coal in their foundry encouraged them to feel this exciting new fuel should be mined, transported and marketed in a developing industrial nation, particularly in the greater Philadelphia area.

Founders of the Lehigh Coal and Navigation Company, Josiah White (left) and Erskine Hazard (right).

Josiah White and George F. Hauto made a tour of exploration of the Lehigh River and the mines at Summit Hill.[4] They returned filled with enthusiasm and faith concerning the practicability of their vision that foresaw development of the mines and transportation of the coal down the Lehigh and Delaware Rivers to Philadelphia markets.

In 1817, the managers of the Lehigh Coal Mine Company executed a 20 year lease of their whole property to Josiah White, George Hauto, and Erskine Hazard.[5] After obtaining the lease, these men applied to the Pennsylvania Legislature for an Act which would authorize them to improve the navigation of the Lehigh River. The Act was duly passed by the Legislators on March 20, 1818.[6] Within five months, The Lehigh Coal Company was formed with capital of $55,000 to construct a road from the river to the mines. The Lehigh Navigation Company was formed to bring coal to market by water.[7] In 1820 the two companies were merged to form the Lehigh Coal and Navigation Company which was formally incorporated in 1822. Events were moving at a rapid pace.

White's pioneering spirit and genius were put to work. The road from the mines at Summit Hill to the Lehigh River, a distance of nine miles, was laid out by use of a level, the first such road of this nature to be constructed in the United States. Horses pulled the first wagons of coal to the loading docks at the Lehigh River.[8] The first boats were arks which were floated down the river. These boats made one trip and then were broken up in the city to be sold for lumber.[9]

Painting of Mauch Chunk, now Jim Thorpe, Pennsylvania, by Herman Herzog, showing the Lehigh Canal in the foreground. *Pennsylvania Canal Society Collection, Canal Museum, Easton*

Original chutes loading the canal boats on the Lehigh Canal at Mauch Chunk. *Pennsylvania Canal Society Collection, Canal Museum, Easton.*

A Lehigh Coal and Navigation Company boat returning to Mauch Chunk after carrying coal to market. *Pennsylvania Canal Society Collection, Canal Museum, Easton.*

A railroad was placed on the horse path and put into operation in 1827.[10] The transportation of coal was by gravity; mules rode down with the coal in their own car, and returned the empty cars to the mines.

Trade in coal soon increased so rapidly that it became evident it would be impossible to continue building a boat for each load of coal to be shipped on the river, so a canal from Mauch Chunk to Easton was begun. Slack water navigation on the Lehigh began in June, 1829; the Delaware division improvements, started 4 months after the Lehigh, were not completed until 1832.[11] The canal required eight dams across the river ranging in height from 5 to 13 feet.

A railroad was also built in 1830 to carry the coal by gravity from Nesquehoning (known as the Rhume Run Railroad) to the canal loading docks at Mauch Chunk.[12]

When the demand for coal accelerated so much that the cars could not be returned to the mines quickly enough by mules, Josiah White decided to carry out his idea of returning the cars to the mines by gravity. To execute this concept, a plane was built from the loading chutes at the Lehigh River to Mt. Pisgah, about 900 feet high. The cars were drawn up this plane by a steam engine; from there they ran by gravity a distance of six miles where they again were raised a distance of 462 feet by another plane to the top of Mt. Jefferson, then continuing another mile by gravity to the mines at Summit Hill. This "back track" was completed in 1845.[13]

"The Switchback"

When mining operations were begun in the Panther Creek Valley the following year,[14] the empty cars descended on a track by gravity and the loaded cars were drawn up a plane similar to that of Mt. Pisgah. The effect was that of a pendulum which swung up and down the sides of the track. The empty car went down the track for a short distance until reaching a place where the road formed into a "Y". There it would go up the hill on the left-hand stem of the Y until it was stopped by the force of gravity. As soon as the car came to a standstill, it began to run down the left-hand stem, and continued on to ascend the road it had come down, to be again stopped by the force of gravity. It would once more start down and, crossing a switch which was closed by a spring, descend on the right-hand stem until it reached another switch, where the same procedure was repeated. It was the peculiar arrangement of the switches, which allowed the cars to move from Summit Hill to what is now Lansford, that gave the railroad system the name that it bore for many years, "The Switchback." It was considered one of the most ingenious feats of engineering work ever accomplished. It attracted many visitors and much attention from the newspapers and magazines of that day.

No. 7 Tunnel, Panther Creek Valley, Pa.

Published for G. A. Ingram

The Hauto Tunnel, completed in 1872. This tunnel connected with the Central Railroad of New Jersey tracks in Hauto Valley. Its use eliminated the need for the gravity railroad or "Switchback Railroad" to transport coal to Mauch Chunk.

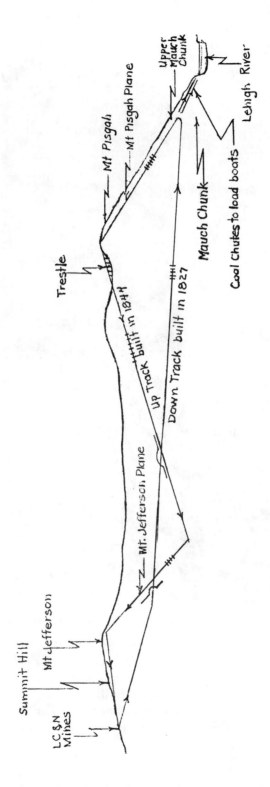

Schematic sketch of Lehigh Coal and Navigation Company gravity railroad.

Base of the Mt. Pisgah Plane of the "Switchback Railroad," Mauch Chunk.

When the Panther Creek Valley was connected with the Lehigh Navigation System by the Nesquehoning Railroad, which passed through the tunnel connecting to Lansford and Hauto in 1870, the original gravity road and the Switchback became useless for hauling coal.[15] The Switchback was leased by the Central Railroad of New Jersey which ran it until it was turned over to a group of Mauch Chunk citizens in 1929. It was used by thousands of visitors yearly to take novel pleasure trips to and from the coal mines. This use was finally abandoned in 1933 and the railroad was sold as scrap to Isaac Weiner of Pottsville for $18,000 in 1938.

In June, 1862, heavy rains caused a great flood in the Upper Lehigh Valley and, as a result, most of the costly canal improvements above Mauch Chunk were destroyed. Damage to the Company's navigation network was so extensive that it was considered useless to try to make repairs between Mauch Chunk and White Haven. The Legislature granted the Company the right to build a railroad from White Haven to Mauch Chunk to connect with the road which had been built from White Haven to Wilkes-Barre. The road, called the Lehigh and Susquehanna Railroad, was soon completed and operated by The Lehigh Coal and Navigation Company. In 1871, the Central Railroad of New Jersey[16] leased the entire road. This lease plays an important role in the remaining corporate affairs of the Company.

The Nesquehoning Valley Railroad had been approved by the Legislature in May, 1861. The road was to begin at the canal landing near Nesquehoning Creek and extend to the headwaters of this stream. After it was built, it carried the coal formerly carried by the gravity road. In 1862, this railroad was continued from Hauto to Tamaqua. It was joined with the Lehigh and Susquehanna Railroad and became part of the Jersey Central system in 1871.[17]

The flood in 1862 convinced the managers that building a railroad was essential.[18] Canals had passed the peak of their importance as a means of inland transportation; the Lehigh Canal had already recorded its largest tonnage of all time seven years earlier, in 1855.

White, the foresighted one, saw the possibilities of getting coal to market from the Susquehanna Valley and points farther west. To do this, he devised a means of transporting coal over the mountains from Wilkes-Barre to White Haven. This he accomplished by means of the Ashley Planes which were still in operation as late as 1948.[19]

The canal was gradually replaced by sections of railroad which finally were united into the Lehigh & Susquehanna Division of the Central Railroad of New Jersey.[20] Later, other sections of railroad were joined together and became the Lehigh and New England Railway. The Lehigh and New England was one of the principal outlets of the anthracite region into the eastern New York and New England markets. In addition to the remarkable developments in canal and railroad modes of transportation, the Company initiated many developments in the mining and preparation of coal.

Josiah White and his partner Erskine Hazard were truly pioneers who founded a great Company through their courage, engineering skill, and vision. The Company continued to grow and develop after their passing under the direction of other capable managers. All were undoubtedly influenced by the pioneering spirit and ingenuity exemplified by its founder. The

Down track of the "Switchback Railroad" alongside of Mauch Chunk Creek at "Two Mile Turn." *Raymond E. Holland Regional and Industrial History Collection.*

Company prospered and built up an enviable record of unbroken dividend payments to its stockholders. Would this Company still be alive and healthy if the managers in the 1940's and 1950's had demonstrated as much ingenuity, imagination and versatility as its founders?

Side by side, the canal and railroads operated close together until the canal was finally abandoned. *Pennsylvania Canal Society Collection, National Canal Museum, Easton.*

THE LEHIGH COAL
AND
NAVIGATION COMPANY

The Lehigh Coal and Navigation Company founded by White and Hazard in 1820 grew and diversified over the years. When the founders merged the Lehigh Coal Company and the Lehigh Navigation Company to form the Lehigh Navigation Coal Company in April 21, 1820, they realized that the Coal and Navigation companies were dependent upon each other and that neither could survive by itself. Two years later the Company was incorporated as The Lehigh Coal and Navigation Company.[21] This interdependence of coal mining and the transportation of the product to the markets continued until the final days of the Company.

After Josiah White relinquished the active management of the Lehigh Coal and Navigation Company in 1841, the Company was headed by the following individuals (although it should be noted that White's partner and friend, Erskine Hazard, and chief engineer, Edwin A. Douglas, continued to play a large role in managing the company's day-to-day affairs until their deaths): John Cox, President, 1841-1842; James Cox, President 1844-1867; E.W. Clark, President 1867-1882; J.S. Harris, President 1882-1893; E.B. Leisenring, President 1893-1894; Calvin Pardee, President 1895-1896; L.A. Riley, President 1896-1907; W.A. Lathrop, President 1907-1912; S.D. Warriner, President 1912-1937; and J.H. Nuelle, President 1937-1938.[22]

Josiah White saw production grow from the modest initial shipments of 1820 to 454,258 net tons in 1850; in 1934 the commercial output of the Lehigh Navigation Coal Company was 3,657,688 tons and the operations had a capacity of 4,500,000.[23] To trace in detail the many changes in the structure and organization of the company which necessarily accompanied this growth would exhaust a volume.

The story of the Lehigh Coal and Navigation Company to be told herein deals primarily with the last turbulent years of the Company from 1938 through 1969 because this period of the Company's history has not been recorded previously. To set the stage, some description of the happenings in the 1920's and 1930's has been included.

First, let's discuss the makeup of the Company as it existed in 1939, one hundred and nineteen years after it was founded.

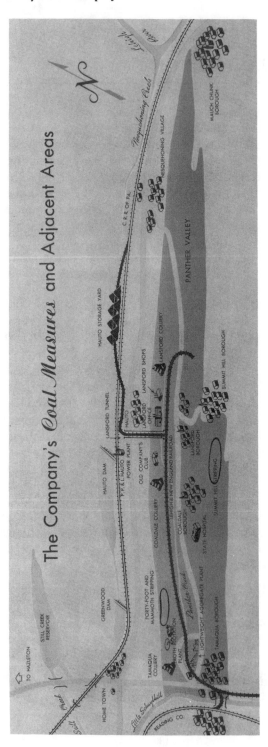

The Lehigh Coal and Navigation Company coal measures and adjacent areas.

1939

THE LEHIGH COAL AND NAVIGATION COMPANY

SECURITIES OWNED, DECEMBER 31, 1939

BONDS AND STOCKS OF SUBSIDIARY AND AFFILIATED COMPANIES:

		Ledger Amounts
3,000 shs.	Capital stock Allentown Iron Company, par value per share $41.60............................	$4,796.93
*4,500 "	Capital stock Allentown Terminal Railroad Company, par value per share $50.................	225,000.00
200 "	Capital stock Blue Ridge Real Estate Company (10 pct. paid), par value per share $50.........	1,000.00
*32,069 "	Capital stock Delaware Division Canal Company of Pennsylvania, par value per share $50.......	1.00
25,000 "	Capital stock Greenwood Corporation, no par value...	1,250,000.00
50,000 "	Capital stock Lehigh Navigation Coal Company Incorporated, no par value..................	3,520,977.06
*135,970 "	Capital stock Lehigh & New England Railroad Company, par value per share $50.............	5,597,699.09
100 "	Capital stock Monroe Water Supply Company (10 pct. paid), par value per share $50.........	500.00
*28,372 "	Capital stock Nesquehoning Valley Railroad Company, par value per share $50.................	1,422,100.00
*7,000 "	Capital stock Panther Valley Water Company, par value per share $100.....................	700,000.00
2,475 "	Capital stock Summit Hill Water Company, par value per share $10..........................	98,805.00
*2,600 "	Capital stock Tresckow Railroad Company, par value per share $50.........................	130,000.00
*10,000 "	Capital stock Wilkes-Barre & Scranton Railway Company, par value per share $50.............	500,000.00
*$377,000	Par value Panther Valley Water Company general mortgage sinking fund 6 pct. gold bonds, due 1943..	358,150.00
500,000	Par value Wilkes-Barre & Scranton Railway Company first mortgage 4½ pct. gold bonds extended to May 1, 1948............................	500,000.00
		$14,309,029.08

OTHER INVESTMENTS:

2,000 shs.	Capital stock Glen Alden Coal Company, no par value......................................	$8,750.00
9,364 "	Capital stock Lehigh & Hudson River Railway Company, par value per share $100...........	627,691.94
500 "	Capital stock Lehigh & Wilkes-Barre Coal Company of New Jersey, no par value.............	1.00
700,000 "	Capital stock National Power & Light Company common, no par value......................	2,859,320.71
	Mortgages..................................	17,230.24
		$3,512,993.89
		$17,822,022.97

* Pledged with the exception of qualifying shares, under the mortgages of The Lehigh Coal and Navigation Company.

Table I. The Lehigh Coal and Navigation Company Securities Owned, December 31, 1939.

(Above) Tamaqua Colliery and froth flotation plant (right), Lehigh Navigation Coal Company, Tamaqua, Pennsylvania. Built 1922; razed June 25, 1958.
(Below) Coaldale Colliery, Lehigh Navigation Coal Company, Coaldale, Pennsylvania. Built 1922; razed 1962.

(Above) Lansford Colliery, Lehigh Navigation Coal Company, Lansford, Pennsylvania. Built 1925; burned 1960.

(Below) Lansford shops of Lehigh Navigation Coal Company employed 350 men at its peak. Initial construction started in 1872. *Courtesy of Clarence Dankel.*

(Above) Nesquehoning Breaker, Lehigh Navigation Coal Company, Nesquehoning, Pennsylvania. Built 1908; razed 1948. *Courtesy of Clarence Dankel.*

(Below) Hauto storage yard enabled Lehigh Navigation Coal Company to stock up to 360,000 tons of coal during periods of low demand. *Courtesy of Clarence Dankel.*

Facilities, Location and Assets of the Lehigh Coal and Navigation Company in the Late 1930's

According to the 1939 Annual Report for the Lehigh Coal and Navigation Company, it was organized as a holding company which operated fifteen subsidiaries. Its corporate headquarters were in Philadelphia, Pennsylvania. Table 1 lists the bonds and stocks of the subsidiary and affiliated companies, as well as other investments of the company in 1939.[24] The Lehigh Coal and Navigation Company and Allied Companies, General Map, shows the location of the Company's facilities, holdings and activities.

The Company's 8,000 acres of coal lands comprised the entire eastern end of the southern anthracite field. The lands began on the east, on the top of Broad Mountain (Mount Pisgah) a half mile from the Lehigh River near Mauch Chunk (now Jim Thorpe), and extended 14 miles to Tamaqua on the Little Schuylkill River. These mining properties were operated by its subsidiary, Lehigh Navigation Coal Company, Inc., with its operating office in Lansford, Pennsylvania.

Two of the Company's railroads, the Lehigh and Susquehanna Railroad, and the Wilkes-Barre-Scranton Railroad, were leased for long terms to the Central Railroad Company of New Jersey for an annual rental of $2,329,015.25.

Main office of the Lehigh Navigation Coal Company, Lansford, Pennsylvania. Built 1872 and enlarged in 1880s. Destroyed by fire in December 1975. Served as headquarters for "Old Company's" Lehigh mining operations. *Courtesy of Clarence Dankel.*

These railroads extended from Scranton to Easton, Pennsylvania. A third railroad, the Lehigh and New England Railroad, extended from Hauto, Pennsylvania, to Campbell Hall, New York, where it connected with the New York, New Haven and Hartford at Maybrook, N.Y.[26] Another branch of this railroad extended to the Lehigh Valley cement plants. Thus this railroad served both the Company's mines in the Panther Valley and the cement mills in the Lehigh Valley. The operating office of this railroad subsidiary was in Bethlehem, Pennsylvania. The photo below shows one of the Lehigh and New England Railroad's steam locomotives crossing The Lehigh River Bridge near Palmerton, Pa.

Lehigh and New England Railroad steam locomotive hauling a train of coal cars across the Lehigh River at Palmerton. *Courtesy of Robert F. Collins.*

The Company owned some 45,000 acres of additional land, including the headwaters of streams with a great potential water supply. Through two subsidiaries, the Panther Valley Water Company and Summit Hill Water Company, the Company met the water requirements of the Panther Valley mines and the domestic water needs of the Panther Valley towns. Much of this land was located in what would become the resort area of the Pocono Mountains.

Another major asset of the Company was its 700,000 shares of National Power and Light Company stock. This stock was acquired when the Hauto Power Plant, designed and built by the Company in 1911, was later sold to the National Power and Light Company, predecessor of the Pennsylvania Power and Light Company.[27]

On the surface, the foregoing group of assets would appear to have been a very sound basis for the building of a long-term future growth. Indeed, from 1938 to 1945, such was the case. However, unknown to anyone and possibly unforeseeable at that time, this mixture of assets was partially responsible for the downfall of the Company, mainly because a deadly conflict of internal interests developed.[28]

The Lehigh and Susquehanna Railroad lease to the Central Railroad brought considerable annual income into the Company treasury. This "easy" money may have had an adverse effect on the management teams over the years not unlike the effect of too much money in the hands of children of well-to-do families. In many cases, such offspring lose initiative, drive and a determination to excel. Something was lacking in the management of the Company in its waning years.

Former Company President H. Louis Thompson further stated, "This conflict of internal interests was between the Coal Company and the Lehigh and New England Railroad. Every effort was made to ship the maximum amount of anthracite over the railroad because of the profit made by this subsidiary. When a strike would occur at the mines, the railroad revenues would shrink because of reduced freight being hauled over the line. The result was a growing chorus of complaints. Officials of the railroad complained to the president of the parent Company. In a very short time, the president of the parent Company would be on the phone with the mines upstate to determine in what way and how expeditiously the miners' grievances could be settled so the men could return to producing coal. As a result, settlements were made at the mining properties which, over the years, permitted many bad practices to develop, which increased costs and impaired productivity. The employees and their union representatives learned from experience that the Coal Company management would avoid taking a hard line on key issues. This situation had a demoralizing effect on the Coal Company management and supervisors."

The People of the Panther Valley

The original settlers in Panther Valley were "Pennsylvania Dutch" and they were among the first miners in the days when most of the coal was quarried. When actual underground mining operations were started, many Welsh, English and Irish immigrants were drawn to the Valley. Many of these men were skilled miners, and although they knew little about the problems of the Pennsylvania anthracite fields, they worked out many of the basic mining methods which were practiced until the mining properties were eventually closed.[29] In the late 1800's and early 1900's, immigrants from the original Austro-Hungarian empire also found their way to the region.

(Above) Dedication ceremonies for the Summit Hill-Lansford-Coaldale Community Pool and Park. *Courtesy of Clarence Dankel.*
(Below) Annual outing of the Old Timers Club (workers of Lehigh Navigation Coal Company with 50 years or more of service). President Evans serving the guests.

Happier times at Old Timers Club outing at Greenwood Lake in Hauto Valley. R.V Happier times at Old Timers Club outing at Greenwood Lake in Hauto Valley. R.V. White, Evan Evans, and J.B. Warriner "singing along."

National (ethnic) churches were established by the various immigrant groups. In the earlier years, these churches were very important support structures until the immigrants learned the language and lifestyle of their adopted country. Although there were various nationalities living closely together, remarkably little friction existed among the residents. Church baseball leagues were very active and helped develop a feeling of tolerance among both participants and fans. Football was almost an obsession with the Valley's residents. Local high school teams played hard, tough ball. Rivalry between towns was strong, and virtually everyone attended the games. The emphasis on sports contributed to the excellence of many players, who were later drafted by various colleges. Local residents always followed with pride the progress of the area's star athletes.

The common interest in sports was a major factor in establishing an esprit de corps among the people. Another common bond was the Union: all the workers for The Lehigh Navigation Coal Company were members of the United Mine Workers of America. The Union was their club, and they were loyal to it.

The local unions for the various collieries, shops and strippings had formed a General Mine Committee. These unions were under District 7 of the International U.M.W.A. The local unions had a long history of "wildcat" or illegal strikes against The Lehigh Navigation Coal Company because they refused to comply with the provisions of the contracts negotiated by their international officers. As a result, the U.M.W.A. officers in Washington had revoked the charter of District 7 and the right of the District to elect its own officers. Provisional officers were appointed by the international officers to lead District 7. Even under the provisional officers, the Panther Valley unions continued to be a rebellious and disorderly group.

According to company records supplied to the author by T.R. Berger, former L.C. & N. Public Relations Director, lost time due to strikes of The Lehigh Navigation Coal Company employees in the late 1940's amounted to 90 percent of all the lost time in the entire anthracite industry although the Company represented only six percent of the industry's production. The record wasn't good.

The mine workers were strongly imbued with the principles of unionism. They wouldn't think of crossing a picket line, and anyone who spoke in favor of some Company action was branded a "Company man," a terrible denunciation.

The miners were capable, hard workers. The nature of their demanding work literally forced them to develop a rugged self respect and self reliance, which made it difficult for them to take orders from anyone, particularly Company "bosses." It is unfortunate that the admirable qualities of the miners could not be fully utilized to work in harmony with the Company supervisors. Both the mine workers and the Company would have benefited greatly.

The miners, their families, and other residents of the Panther Valley were friendly, helpful people when it came to anything not connected with the Company. They enjoyed each other's company and were proud of their respective communities, churches, and many social clubs. No one wanted to leave the "Coal Region" to work elsewhere. When the mines eventually shut down, most of the men commuted to jobs in Allentown, Bethlehem, Easton and New Jersey.

"The Old Company" as a Corporate Citizen

Lansford, Coaldale, Tamaqua, Summit Hill, and Nesquehoning were "Company towns." Almost everyone in the area worked for The Lehigh Navigation Coal Company or provided services for those who did. In the early years, the Company built houses for the employees. Rent charged for the homes was extremely low. A company store, later Bright's Department Store, sold to the miners on "the cuff." On payday, the Company deducted the amount owed from the miners' pay.

Since the Company owned almost all the land and all the mineral rights, as well as numerous homes and other buildings at the various area towns, it was by far the largest taxpayer. In Nesquehoning, for example, 60 percent of all taxes were paid by The Lehigh Navigation Coal Company.

The Company was generous in its help to local governments, schools, churches, and associations of all kinds. It was a "soft touch" for almost every community activity, particularly because so many officials and supervisors were active in many civic and social activities. The real estate department, with its own maintenance crew, was called upon to do much public service work. Company services were used quite freely by many of the bosses and employees. One observant local resident likened the Company to a "sow pig with thousands of piglets sucking away on its tits."

The Company was generous in making gifts for the community projects. It provided space for a youth center in Lansford and donated a former mule stable in Nesquehoning to become a recreational facility in that town. President Evans arranged to have the Company donate land and $50,000 as seed money for the construction of a community pool and park to serve the Summit Hill, Lansford, and Coaldale communities. A Tri-Borough Commission was established to operate the recreational area. The pool and park were dedicated on July 4, 1944.

Evans was farsighted in recognizing that the Company could not provide jobs for all the Panther Valley workers. He was instrumental in organizing the Panther Valley Industrial Development Commission, later to become the Carbon-Schuylkill Industrial Development Commission. The commission was initially headed by the Company's arch enemy, James H. Gildea, editor of *The Coaldale Observer*. The Bundy Company was the first company to establish a plant in Hometown as a result of the commission's work. A number of other corporations were encouraged to locate plants in the Valley during the ensuing years.

Company foremen and officials were active in the various town school boards, governments and other community activities. In earlier years, the Company supervisors controlled local politics because of their influence on the workers. Republican candidates usually won elections. The strong Company influence on local politics when the Company was "supreme" eventually had an adverse effect on the relations between the Company and its workers.

However, the paternalistic attitude of the Company was beneficial to the local residents. The strong tax base provided by the Company made it possible for the Panther Valley towns to have better streets and community services than other mining areas in the anthracite region. The Lehigh Navigation Coal Company was a good corporate citizen.

General Industry Conditions in the 1920's

Great prosperity prevailed in the country during the decade from 1920 to 1930. The anthracite industry reached a post World War I production peak of 77,603,000 tons in 1923.[30] However, from that time on, the industry started to lose markets from competitive fuels made available to the public. Oil, gas, and prepared bituminous (consisting of 60 percent anthracite and 40 percent bituminous) were finding their way into the anthracite consuming area.

Prolonged strikes at the mines at critical times, such as in the middle of the winter, also caused shortages. Anthracite customers were greatly dissatisfied and discomforted; they began to feel that it was not a dependable fuel. Furthermore, dusty coal bins in the cellar and the inconvenience of tending a furnace and removing ashes were disadvantages that could now be eliminated by using gas or oil heat. Unsettled labor conditions pressured the public into converting to these other forms of fuel, but increased convenience was undeniably another impelling factor. This market attrition grew, and anthracite production declined to 64,203,000 tons in 1929, a year of great prosperity for most other industries.[31]

Thus, it can be seen that the industry was facing difficulties in holding its markets even before the "Great Depression" in 1929. During the depression years, the anthracite market continued to shrink. Production dropped to 41,780,000 tons in 1933, almost half of what it was 10 years earlier.[32]

Samuel D. Warriner, President of Lehigh Coal and Navigation Company 1912-1937; Manager 1912-1941.

The Lehigh Coal and Navigation Company was headed by a veteran mining man, S.D. Warriner, during this period. S.D. Warriner was one of the most influential men in the anthracite industry for many years. After graduating from Amherst, he received his Engineer of Mines degree at Lehigh University in 1890. He had a varied career. He mined iron ore in Virginia, served as a mechanical engineer for the Lehigh Valley Coal Co., (a subsidiary of the Lehigh Valley Railroad until 1923) and served as engineer in charge of development work for the Calumet and Hecla copper mines in Michigan. He then returned to Wilkes-Barre to succeed Mr. W.A. Lathrop as General Manager of the Lehigh Valley Coal Company.

Because of his excellent background and experience in the mining field and his outstanding reputation in the anthracite industry, he was elected President of The Lehigh Coal and Navigation Company in 1912 and served in that capacity for 25 years, the longest term of any president.

Sam Warriner's father was a country minister. However, this religious influence didn't seem to soften him when it came to handling the unions and his stance on many labor grievances which he heard as a member of the Anthracite Conciliation Board. He was strongly opposed to the "check off" system demanded by the miners, and insisted that welfare work must be cooperative on the part of the company and the men: "Other plans rob the men of self respect and create paupers."

Sam D. Warriner provided strong leadership for the Lehigh Coal and Navigation Company, as well as for the industry. This strong, silent, intense man left his mark on the Company and the industry.[33]

"The Old Company" in the 1930's

In 1930, S.D. Warriner, President of The Lehigh Coal and Navigation Company, elevated his cousin, Jessy B. Warriner, to President of The Lehigh Navigation Coal Company. J.B. Warriner was born in Montrose, Susquehanna County, Pennsylvania, in 1883. His father was an Episcopal minister. He received his Bachelor of Science degree in Mining Engineering in 1905 from Pennsylvania State College. He worked as a mining engineer for the Lehigh Valley Coal Company, the Delaware, Lackawanna and Western (Glen Alden Coal Company), and then for the Northwest Improvement Company in the state of Washington. His first supervisory position was as superintendent of the mine of Denny, Renton Clay and Coal Company of Seattle, Washington.

In 1913, Warriner moved back East to accept a job as mining engineer and operator for G.B. Markle Company (later the Jeddo Highland Company). Shortly after that, he started his long career with The Lehigh Navigation Coal Company as Chief Engineer. He advanced to the position of

General Manager in 1922 and Vice President in 1926. Warriner was elected President in 1930.

In addition to his company duties, he served as a member of the Anthracite Board of Conciliation where he gained the appreciation not only of the operators, but also of organized labor.

Warriner belonged to the American Institute of Mining and Metallurgical Engineers and the Coal Mining Institute of America, and served as director of the American Mining Congress. He was also an avid fisherman and a capable golfer.[34]

Jessy B. Warriner, President of Lehigh Navigation Coal Company 1930-1946; Manager 1942-1948.

In line with the rest of the anthracite industry, production of coal from the Old Company's lands dropped from a peak of 5,000,000 tons annually to 2,223,886 tons in 1937.

Losses by the Lehigh Navigation Coal Company increased as the volume of production declined. Reduced production meant less operating time at the mining operations. Curtailed working time for the miners meant labor unrest, which meant strikes and further inefficiencies.

The financial record of the Lehigh Navigation Coal Company during this period was one of constant losses resulting from diminished business and continuing high operating costs. The Company earned no profit from 1927 to 1940. In the period from 1930 to 1940, its accumulated loss was $9,865,587. This loss amounted to 34.6 cents on each ton of coal produced.

In order to keep the coal properties in repair and in operation during these years, the parent company had to forego the collection of moneys owed to it for power, royalties and other items, in addition to making outright loans and advances. At the end of 1940, the Coal Company owed the parent company a total of $5,545,603. Moreover, the parent company paid the taxes on the properties operated by the "Old Company."[35]

Because of the decline in markets for anthracite coal starting in the 1920's and continuing through the 1930's, many mines were closed by the "main line" coal producing companies throughout Northeastern Pennsylvania. When the miners were put out of work, many entered upon the coal lands of the coal companies and opened up operations of their own. These operations were called "bootleg mines." The legitimate operators sought relief from State and local law enforcement officers to no avail. The "bootleg" miners had political clout. They formed their own associations. Anarchy continued unabated with the result that 3,600,000 tons were mined and marketed by bootleggers in 1935, representing one month per year of business for the legitimate miners.[36]

Because production from bootleg and legitimate mines exceeded the market for coal during these years, prices of coal were depressed. Furthermore, cost of operating the mines at reduced and arbitrary working schedules exceeded the prices received for the product. Accordingly, losses at the mines mounted.

The Big Three: W.H. Edwards, President, Lehigh and New England Railroad; R.V. White, President, Lehigh Coal and Navigation Company; J.B. Warriner, President, Lehigh Navigation Coal Company.

During this period, the coal companies were negotiating to secure lower wage scales. Direct negotiations with the U.M.W.A. for relief failed when the Secretary of Labor requested a postponement, pending President Roosevelt's recovery program.

Labor troubles became pronounced by 1933 at the Old Company's mines in the Panther Valley with agitation for exact equalization of working time among collieries regardless of cost. Encouraged by local politicians and even members of the clergy, strikes were called by the mine workers demanding this concession of management. Unrest at the mines became more acute in the next few years. Early in 1935, several strikes occurred when the Union demanded that stripping operations (surface mining) be curtailed to make it possible to operate the deep mines for a greater number of days. Also during 1935, a so-called Equalization Committee, made up of employees of all Panther Valley operations, was organized, and functioned throughout the year. This body was not sanctioned by the Company's agreement with the United Mine Workers of America, but it asserted the right to dictate to the Company the order and schedule by which collieries operated. In addition, the committee endeavored successfully to restrict strippings and other operations. These activities added to the costs of operations and handicapped sales efforts by frequently making it impossible to fill special orders.[37]

One of the leaders of the movement to force the Old Company into an Equalization Agreement was the editor of *The Coaldale Observer*, James H. Gildea, best known as "Casey" Gildea. He played a dominant role in promoting the concept by constantly publishing supporting articles in his paper and promoting a high personal visibility at parades and mass meetings of the miners. He was elected to two terms in the United States Congress, largely through such publicity.

The Company management finally capitulated to the demands of the miners on equalization. A two year agreement negotiated with the U.M.W.A. officials, effective May 7, 1936, provided for parceling out working time at the collieries and a complete "check off" system for collection of Union dues by the companies. This contract also provided for a seven hour day instead of an eight hour day, but at the same wages.[38] All this occurred at a juncture when anthracite was in a steady downturn.

The Equalization Agreement was significant and would have far reaching consequences for the future of the Company. Its interpretation and the shortened work day, plus complicated wage formulas, were mainly responsible in later years for the complete closing of the mines by the parent Company. The miners won the battle for the short term, but lost the war over the long term.

This 1936 contract, in establishing reduced working hours without a corresponding adjustment in wage rates, created higher costs of production. Also, the automatic "check off" of Union dues further strengthened the Union by making collection mandatory by the Company.

Samuel D. Warriner, a veteran and highly esteemed mining man, was President of The Lehigh Coal and Navigation Company during these turbulent years. It is no wonder that he requested a relief of some of his duties. On April 1, 1937, the Board acceded to his request and elected him Chairman of the Board. He was succeeded as President by J.H. Nuelle.[39] Nuelle soon left to become President of the Delaware and Hudson Railroad. The financial strength of The Lehigh Coal and Navigation Company made it possible to continue to pay dividends to its stockholders throughout the depression years, despite the heavy losses of the Coal Company subsidiary. A consolidated loss of $306,510 in 1937 forced the Board of Managers to reduce the dividend to 30 cents per share from 60 cents in the previous year. The Board decided to do something about it.

Several large stockholders of the Company asked Robert V. White, a member of the Board of Managers, to accept the presidency and to terminate his business connections in New York. As an inducement to White these shareholders executed options for a total of 20,000 shares of common stock of the Company, good until September 1, 1941. Mr. White agreed to this arrangement and was made President of the Company on September 16, 1938. So began the "White Era."

THE WHITE YEARS (1938-1953)

Robert V. White, President of Lehigh Coal and Navigation Company, 1938-1953.

Mr. Robert Vose White was born in Boston, Massachusetts, on December 24, 1886. He received his A.B. from Harvard in 1909 and his L.L.B. in 1912. He practiced law with Cravath, Henderson & de-Gersdorff in the years 1912-1918. He entered the office of J. & W. Seligman & Co. in 1918 and became a partner in 1920. In 1937, he was made a partner of Jackson & Curtis. He was a governor of the New York Stock Exchange from 1935 to 1938, and chairman of the Stock List Committee in 1938.[40] Such was the background of the new President of The Lehigh Coal and Navigation Company.

Mr. White, 51, an aristocratic looking executive, took over the reins of this famous Old Company in April, 1938. He had a full head of white hair and wore pince-nez which accentuated his Ivy League appearance. Tall and slim, his posture was always erect, and he was in good physical condition. While at Harvard, he was a member of the football team. He was also an avid skier.

One of Mr. White's first moves as president of the parent Company was to encourage the coal mining subsidiary, the Lehigh Navigation Coal Company, to bring new people into the mining company. J.B. Warriner, president of the Coal Company, had held that position since 1930. He was a graduate mining engineer with excellent mining background and experience. Most of the Coal Company staff and supervisory force were men who had come up through the ranks and had developed local ties which made it difficult for them to administer the Lehigh Navigation Coal Company from an unbiased viewpoint.

Warriner was happy to accede to White's suggestion of recruiting trained engineers into the Company. As a result, Charles S. Kuebler and Albert Beckwith, mining engineering graduates from Lafayette College, were hired. These men were followed by John McGrath, mining engineer from Columbia University, and W. Julian Parton, who had received his master's degree in mining engineering at the University of Washington in June, 1939.

Meanwhile, White, from his Philadelphia base, intensified his recruitment of the "new look" group by hiring personnel for the parent Company organization. In 1941, he hired an assistant secretary, Glenn O. Kidd, a young attorney who had worked for the Interstate Commerce Commission in Washington, D.C.; J.S. Hanks, who had a banking and accounting background, was hired in 1943 as assistant comptroller. Others were Vernon Boyles, assistant treasurer; Robert L. Boyd, assistant secretary (1945); John C. Bolinger, Jr., assistant secretary (1947); John Rutledge, assistant comptroller (1949); H. Louis Thompson, comptroller (1951); Harry W. Siefert, assistant comptroller (1951); and John S. Gates, secretary and assistant treasurer (1952). F.M. Thayer, Jr., as assistant secretary, and C. Peter McColough, as sales manager of The Lehigh Navigation Coal Company, were both hired in the fateful year of 1953. Not a few were graduates of big name business schools most notably White's alma mater, Harvard.

With so much "new blood" being infused into the Company, it was inevitable that many changes in organization would occur during the White years. The new employees were well trained, skilled, and capable, as can be proven by what they were able to accomplish after they left the employ of the Company. H. Louis Thompson, an accountant by training and a former F.B.I. person, was to play a dominant role in the latter years of the Company as its President. Kidd and Parton succeeded to the presidency of the Coal Company before it was closed and ceased operations. After leaving the Company, Kidd became a successful New York business consultant specializing in corporate mergers. Bolinger became President of Mississippi River Fuel Corporation. McColough went to Rochester, New York and later to become Chairman of the Board of Xerox Corporation. Rutledge moved up to vice president of Xerox. Parton became President of The General Crushed Stone Company and a Vice President of Koppers Company. Clearly, White had recruited a talented group of people to help him turn around the fortunes of The Lehigh Coal and Navigation Company and its subsidiaries.

In addition to his program of recruiting capable personnel, White employed a management consultant who had worked with him when he was on Wall Street. Henry C. (Pop) Schwable was an expert on planning, forecasting, and budgeting. He spent most of his time between the parent company offices and the Coal Company offices in Lansford working with and indoctrinating the staff on modern methods of management. He took particular interest in the newer recruits of the Company, apparently because he felt that they would be more open to the new concepts he wanted to instill in the management team. His teachings and guidance helped to develop the management skills of the people with whom he came in contact.

Labor troubles and poor employee attitudes were responsible for the low productivity and the high costs at the mining operations in the Panther

Valley. In 1946 White employed Lewis and Gilman, a new advertising and marketing agency headquartered in Philadelphia, to establish programs for improving employee relations and to handle the advertising programs aimed at promoting the sale of anthracite. Also, this agency was to work with financial editors in metropolitan areas to improve the Company's public relations image. The budget for these programs was in excess of $500,000 annually.[41]

Weekly advertising messages were placed in the *Allentown Morning Call*, *Lansford Record*, *Tamaqua Courier*, *Coaldale Observer*, and the *Mauch Chunk Times News*. Letters and annual reports were mailed to employees and civic leaders. Cause of strikes was publicized immediately, and the Company's side of the dispute delineated carefully. Lost earnings of the miners when the strikes occurred were detailed.

In order to best carry out the employee relations program in the Panther Valley, Lewis and Gilman assigned a former local resident, T.R. Berger, to manage the public relations programs. Berger had been an editor of the *Allentown Morning Call* and knew the territory well.

A far-reaching sales promotion program was carried on by use of the radio. The pioneering "Lehigh Weatherman" broadcast daily weather forecasts in the major market areas of the Company, principally in Pennsylvania, New York, New Jersey, and the New England states. The revolutionary notion that a program could use four minutes to talk about the weather caught on.

These were some of the things that the new president of The Lehigh Coal and Navigation did when he took over the management of the Company.[42]

Pre-War Years (1938-1941)

The financial position of The Lehigh Coal and Navigation Company and subsidiary companies in the first year of White's presidency is shown by the consolidated balance sheet, December 31, 1938 in Table II. The balance sheet shows total assets of the Company of $92,768,623 and a net worth of $42,728,868. Funded debt at this time was $32,123,475.[43]

The Company operated at a net loss of $46,815 in 1938 as compared with a net loss of $306,510 in 1937. Despite the loss, the Company paid 10 cents per share dividend on its 1,929,127 shares of common stock outstanding.

The Lehigh Navigation Coal Company, Inc., was the major source of difficulty, since this subsidiary showed a loss of $1,050,548 for the year. The Coal Company sold 2,424,869 tons of coal that year, a decrease of 19,061 tons as compared with the previous year.

Borrowings from the Lehigh Coal and Navigation Company to meet current expenses increased from $325,000 to $2,625,000. The Coal Company thus continued to be a heavy drain on its parent Company.

14   *The Lehigh Coal and Navigation Company*

1938

THE LEHIGH COAL AND NAVIGATION
CONSOLIDATED BALANCE

ASSETS:

CURRENT ASSETS:
Cash	$2,435,377.21	
Customers' notes and accounts receivable, less reserve for accounts doubtful of collection	2,636,133.36	
Coal (in storage and in transit)	1,092,047.46	
Materials and supplies	764,848.70	
Sundry debtors	433,518.57	
Working funds	63,048.00	$7,424,973.30

INVESTMENTS:
Free (Note 1)	$3,698,891.07	
Pledged	222,500.00	3,921,391.07

FIXED ASSETS:
Coal lands, mining and marketing property	$29,900,369.07	
Canal property (Note 2)	3,736,239.78	
Railroad property	40,874,701.11	
Water property	3,289,846.24	
Real estate	946,099.36	78,747,255.56

DEFERRED ASSETS:
Stripping expenses deferred	$3,127,164.08	
Less reserve for stripping expenses	1,515,356.08	
	1,611,808.00	
Deferred and suspended accounts	1,054,975.88	2,666,783.88

SINKING FUNDS, CASH	8,219.43
	$92,768,623.29

NOTES:
1. Of these investments, certain securities aggregating $2,868,070.71 amount to $5,785,000, based on market quotations at December 31, 1938. The remaining securities included herein are not quoted.
2. Canal property includes $1,619,870.64 representing the ledger value of the canal property of the Delaware Division Canal Company, as carried on its books of account. The capital stock of the Canal Company has a par value of $1,603,450 and is carried in the investment account of The Lehigh Coal and Navigation Company at $281,910.51. The difference of $1,321,539.49 is included in capital surplus.

15   *The Lehigh Coal and Navigation Company*

1938

COMPANY AND SUBSIDIARY COMPANIES
SHEET, DECEMBER 31, 1938   LIABILITIES:

CURRENT LIABILITIES:
Audited vouchers and pay rolls	$1,353,387.00	
Sundry creditors	99,846.85	
Accrued taxes	691,636.73	
Matured and accrued interest	616,601.18	$2,761,471.76

FUNDED DEBT:
Outstanding		$46,421,474.72	
Less treasury bonds:			
Pledged	$400,000.00		
Unpledged	13,898,000.00	14,298,000.00	
			32,123,474.72

(Notes, equipment obligations and payments to sinking funds due in 1939 aggregate $398,074.17)

MORTGAGES PAYABLE (Installments aggregating $6,500 due in 1939)	254,125.00

DEFERRED LIABILITIES:
Compensation claims determined	$180,163.42	
Deferred and suspended accounts	331,762.14	511,925.56

RESERVE ACCOUNTS:
Depletion, since December 1, 1937	$58,410.75	
Depreciation	13,864,666.17	
Workmen's compensation insurance	73,120.45	
Taxes	344,367.16	14,340,564.53

MINORITY INTERESTS	48,193.34
	$50,039,754.91

CAPITAL STOCK AND SURPLUS:

Capital stock, authorized 3,000,000 shares:
Issued:
1,929,100 shares without par value		
9 shares, $50 par value		$32,152,116.67
(Exchangeable for 27 shares without par value stock)		

SURPLUS:
Capital—represents net excess of par values of stocks and bonds of subsidiary companies over amount at which carried on the books of the parent company, net of goodwill of certain other subsidiaries eliminated (Note 2)	$2,202,377.88	
Appropriated under rulings of Interstate Commerce Commission	51,621.56	
Appropriated—Sinking fund reserves	41,999.31	
Funded debt retired through income and surplus	208,452.25	
Earned surplus, as annexed (Note 1)	8,072,300.71	
	10,576,751.71	42,728,868.38

	$92,768,623.29

Table II. Lehigh Coal & Navigation Co. and Subsidiary Companies Consolidated Balance Sheet, 12/31/1938.

The annual report of the Company for 1938 stated that mild weather prevailed throughout the year, resulting in a 14 percent reduction in production for the anthracite industry. Because the limited sale of anthracite resulted in much idle time at the mining operations, the Coal Company negotiated a lease of one of its four mining operations, Nesquehoning Colliery, for five years to the Edison Anthracite Company of Scranton, Pennsylvania. It was hoped that by this lease new markets for coal mined at Nesquehoning would increase total production of the Company's mines and reduce operating costs.

As usual, the railroads contributed substantial income to the parent company in 1938. Rentals from the Central Railroad Company of New Jersey for the Lehigh and Susquehanna lease amounted to $2,279,096 and $66,606 for the Wilkes-Barre and Scranton Railroad. The Lehigh and New England Railroad made a profit of $337,798, a decrease of $45,145 from the previous year.

Other sources of income were not too significant to the whole Company picture. The Lehigh Canal operations lost $28,270 and the water companies made a combined net income of $34,798.

This was the situation in The Lehigh Coal and Navigation Company in 1938. Due to its heavy losses, the Coal Company was using up most of the income from other sources. This had been, and was, the main problem facing White as he finished his first year as president of the Lehigh Coal and Navigation Company. He had been able to reduce the magnitude of the loss compared with the previous year, but that was all.

White was able to show further improvements in the operating results of the Lehigh Coal and Navigation Company in 1939. Total production from the Coal Company properties increased, including that from the Nesquehoning Colliery which had been leased to Edison the previous year. With this improvement, consolidated profits were $18,674, and dividends of 30 cents per share were paid the shareholders, compared with 10 cents per share in each of the last two years.[44]

Because of the brighter outlook for the Company, in 1940 the Coal Company was permitted to spend $650,000 on capital improvements aimed at reducing operating costs. A drainage tunnel was driven at Tamaqua to reduce pumping water from the mines. Further consolidation of the Tamaqua and Greenwood Collieries was accomplished. Electrification of Coaldale Colliery was completed by installing electric mine hoists to replace steam driven hoists.[45]

Evan Evans had been acting general superintendent of the Coal Company and on December 26, 1940, he was named vice president and general manager. He would report to President Jesse B. Warriner.[46]

The year 1941 was a good one for the Company. The Coal Company subsidiary showed a profit for the first time since 1927 as production increased to 3,250,950 tons, to 6.96 percent. Prices were holding firm under the stabilization plan. Coal Company profits were $412,178.[47] Officials of the parent company and the Coal Company were elated with these results. White, undoubtedly, also was happy with results and felt that the Company was making a turnabout under his leadership.

With Coal Company losses halted and income from the Lehigh and New England Railroad profits up to $1,043,103 because of increased tonnage of coal being hauled, abetted by the regular rental income from the Central Railroad of New Jersey for the Lehigh and Susquehanna lease, net profits for the parent company increased to $1,805,057.[48] The dividends for common stock increased to 65 cents a share, the highest paid since 1932.

Despite all the good news, however, the bankruptcy in 1940 of the Central Railroad of New Jersey was an ominous development. Trustees of the Central Railroad of New Jersey had been appointed under Section 77 of the Bankruptcy Act. Under Orders of the Court, the Trustees had paid $1,916,950 for rentals for the Lehigh and Susquehanna Railroad and branches in 1940 -$370,000 less than the yearly stipulated rental. The Court gave the Trustees the right to affirm or disaffirm the Lehigh and Susquehanna lease by May 1, 1941. The Company had prepared contingency plans to recover the Lehigh and Susquehanna Railroad and the Wilkes-Barre and Scranton Railroad for operations in the event of rejection of the leases by the Trustees.[49] The Company was girding for a long legal battle to protect its earnings from these important railroad leases. This major source of income for the Company had enabled it to withstand hard times. An enviable and unbroken record of dividend payments to the stockholders since 1871, when the lease of these railroad properties was initially made, had been maintained for 70 years.

On May 6, 1941, the Court extended the time for election by the Trustees until further Order of the Court, and at the same time directed the Trustees to pay forthwith to The Lehigh Coal and Navigation Company the $368,222 in back rentals being withheld by the Central Railroad of New Jersey. On June 30, 1941, the Court further entered into an Order directing that until the Trustees elected to affirm or disaffirm the leases, they should pay the full amounts of rentals stipulated under the leases as accrued.

With the Coal Company operating at a profit and with difficulties with the Central Railroad of New Jersey leases temporarily resolved, White's standing with the Board of Managers and the stockholders of the Company was high. In the summer of 1941, some of the stockholders discussed with White his continuance as president and the renewal of the stock options which had been granted to him in 1938 to induce him to take over the presidency of the Company. However, in September, 1941, the Board of Managers believed that the stock options should be offered by the Company and not on the part of certain stockholders. Accordingly, the board authorized an agreement with White under which the Company purchased in the market 20,000 shares of its common stock for his account. White agreed to pay all costs to the Company on or before September 25, 1944. The stock was held as collateral and interest was charged at three percent per annum.[50]

The War Years (1941-1945)

On December 7, 1941, "the day that will live in infamy," America was attacked by Japan and projected the country headlong into World War II. The country had been mobilizing slowly during the pre-war years in order to provide arms and help to England and Western Europe. Now war time mobilization was in full force. Demand for anthracite had been increasing prior to this country's entry into the war with higher shipments to Canada and increased industrial activities at home. War-time demand for anthracite resulted in further increases in production and continuing profitable operation of the coal properties. Greater shipments of coal over the Lehigh and New England Railroad further increased the subsidiary's profits.

The 1942 annual report stated: "Lehigh Navigation Coal Company Incorporated continued its program of development and improvement of the company's coal properties. This resulted in operating economies and contributed to the Coal Company's ability to meet the increased demand for anthracite created by the war emergency."

Major improvements completed or under way included consolidation of Greenwood and Tamaqua collieries, reopening of the old Rahn mine and erection of a plant to recover by means of a froth flotation process the very fine sizes of anthracite, hitherto largely wasted.[51] The company was "gearing up" its productive capability.

Substantial investments were likewise made in the Lehigh and New England Railroad during this time. Orders had been placed in 1941 for 215 hatchway roof, hopper bottom steel cars to enable the company to better handle its cement tonnage. Also, 300 steel hopper cars were ordered. Final shipment of these cars was made in 1942. Four used Mikado-type locomotives were also purchased.

The acquisition of these additional freight cars enabled the company to lease out its surplus rolling stock at a substantial profit. These new cars also enabled the railroad to render better service to its shippers in spite of the heavy demands.[52]

The impact of the War was felt in the Company, with 558 employees answering the call in 1942 to serve in the armed forces. A payroll deduction plan was instituted for employees of The Lehigh Coal and Navigation Company and its subsidiaries for purchase of "War Bonds."

Operating results in 1942 were about the same as the previous year; a dividend of 65 cents per share was again paid to the stockholders.

S.D. Warriner died on April 3, 1942. He had been elected president on July 1, 1912, and served in that capacity until April 1, 1937, a continuous term of 25 years, longest in the history of the company. He continued on as a manager and chairman until his death.

The Lake Harmony development in the Western Pocono region of Carbon County continued to grow after its inception in March, 1937. Split Rock Lodge and Club on Lake Harmony, a unique mountain resort, opened in February, 1942. Land leased to Split Rock Club included thousands of acres of land and over 50 miles of stream. This Club offered its members exceptional fishing and hunting privileges; the Lodge was a popular resort from the very onset and grew rapidly.

Original Split Rock Lodge at Lake Harmony, Pennsylvania. *Courtesy of George Harvan.*

The full impact of the War was felt by the Company in 1943. Production from the coal properties hit new highs. A six-day week was instituted. The miners cooperated by working more days per month than they had done for many years. Later in the year, however, because of difficulties in negotiating a new contract with John L. Lewis and his U.M.W.A., the Solid Fuels Administration took possession of all anthracite mines in May, 1943. This arrangement terminated October 13, 1943, but labor difficulties again forced possession one month later.

Because of high production and shipments of anthracite, profits from the Coal Company and the Lehigh and New England Railroad brought a new high in net earnings of The Lehigh Coal and Navigation Company. The dividend was raised to 90 cents per share in 1943. The Company had paid a dividend every year since 1881, or for 63 years.[53] Quite a record!

The annual report for 1943 describes the feelings of management well: "Prospects are that very large production (anthracite) will be necessary for some years, both to meet the war programs and also for the peace years to follow. Of course, it is probable that the industry after the war will experience again for a time many of the problems of the last two decades, but there seems to be good reason for believing that a number of these problems will be solved."[54]

It is evident that White's management strategies were clouded by the high demand for anthracite caused by the war. He also felt that the industry could maintain the markets created by war activity. During this period, the Company sponsored activities of the Anthracite Institute which developed and promoted ways to burn anthracite in a more convenient manner. There was great hope that Automatic Coal Burner Company and Furnaceman, Inc., would be successful in promoting automatic and semi-automatic anthracite-burning equipment. Its efforts were backed by the Company and other anthracite interests. Furnaceman, Inc., also performed services such as furnace tending and ash disposal.

The greatest impact of the War was experienced in 1944 when production of coal, shipments and profits all reached new highs. Consolidated profits were $3,381,093. The dividend went up again — to $1.00 per share. The total of dividends paid was the largest since 1931.[55]

The Coal Company produced 4,664,602 tons in 1944, 14 percent more than in 1943, and made a profit of $1,253,667.[56] Output had increased by 93 percent since 1939 when production of defense materials for the United States and its allies demanded increasing supplies of fuel. The principal concern was how to hold the gains of war-time years.

Anthracite Industries at Primos, Pennsylvania, announced the development of a new coal-burning device called the Anthra-Tube. It was hoped this development would aid in keeping the new markets for anthracite. The Coal Company continued to expand its mining operations with the 250-foot extension of the shaft at Tamaqua Colliery to a new fifth level. A 1,300-foot tunnel was driven to develop north-side workings and an air-chute from the new fifth level to the fourth level was completed. Nos. 4 and 5 mines at Lansford were being readied for production after having been flooded in the damaging floods of 1928. After these floods, the mines were allowed to fill to drown out four mine fires. The largest motor-driven hoist in the anthracite region was installed at Coaldale Colliery.

Government possession and control of the coal mines terminated by order of the Secretary of the Interior on June 21, 1944.

Company employees in the Army, Navy, and auxiliary forces reached 1,200 by the end of 1944.

A significant development in the year was the amicable settlement of all differences between the Central Railroad of New Jersey and the Lehigh Coal and Navigation Company, Inc. The Central Railroad of New Jersey was given the right to sublease the Lehigh and Susquehanna Railroad and its branches to the Central Railroad of Pennsylvania, its wholly owned subsidiary, but the Central Railroad of New Jersey remained liable under the lease.[57]

The last year of the War, 1945, was a disappointing one for Lehigh Coal and Navigation Company. In the face of this, White stated in the annual report for 1945 that consolidated earnings for the year bore very little relation to the actual operating profits. He attributed this to adjustments in connection with charges for amortization of railroad cars purchased during the war period as permitted by the Federal Income Tax Law and expenses in connection with refinancing, which resulted in income reduced by $1,101,705 and consequent reduced taxes. He reported that net income in 1945 was $1,703,379 as compared with $3,381,093 in 1944.[58]

The Coal Company, now known as Lehigh Navigation Coal Company Inc., saw profits take a "nose dive" from $1,253,667 in 1944 to $447,733 in 1945. Production was 3,391,488 tons, a drop of approximately 700,000 tons in one year. This reduction was explained to be "due in part to the anthracite strike, the severe cold weather in the early winter months, the heavy rainfall in the summer months, and other interruptions to operations." The poorer results of the coal properties were not viewed with much concern despite the still relatively high production and sales during this last year of the War. Management was still basking in the rosy glow of the prosperous war years; it could not foresee that the decline in the coal business was going to continue at a rapid rate in the coming years.

A froth flotation plant was under construction at the Tamaqua Colliery. This plant was designed to reclaim fine particles of coal formerly lost in the colliery wash water. It not only recovered and beneficiated a former waste product, but also enabled the Company to meet the requirements of recent legislation imposing new limits on stream pollution.

From the time he became president in 1938, White adopted a policy of reducing the funded debt of the company in the hands of the public. The bonded debt was $13,000,000 at the end of 1945, a reduction of 43 percent since 1938. Lehigh and New England Railroad Company's bonded debt was $4,000,000 in 1945, a reduction of 45 percent since 1938. Because of this policy, annual interest costs were substantially reduced.[59]

Froth Flotation Plant at Tamaqua Colliery designed to recover fine sizes of anthracite and prevent stream pollution.

The group inspecting the 40-foot and Mammoth stripping excavation to eventually total 44,000,000 cubic yards of material removed from an area 3,000 feet long, 1,600 feet wide, and 650 feet deep.

The Lehigh Coal and Navigation Company submitted to a commission appointed by the Philadelphia mayor a plan under which the upper Lehigh River and its tributaries could be utilized for supplying pure mountain water in sufficient quantities and at a reasonable cost to the city.[60] A decision was expected to be made between water from the upper Lehigh and the upper Delaware for city use. The Lehigh Coal and Navigation Company would benefit handsomely if it were able to sell its water rights in the Lehigh and the Pocono Mountains to the City of Philadelphia.

The Post-War Years (1946-1953)

The remaining years of White's presidency of The Lehigh Coal and Navigation Company can be divided into two periods: the "years of warning," 1945-1948, and the "disastrous years," 1949-1953.

Anthracite industry sales held up fairly well during the first three years following the end of World War II. However, The Lehigh Navigation Coal Company had increased its capacity during the War years by 93 percent. This was a good thing since it enabled the Company to best meet the country's war-stimulated demand when a shortage occurred for competitive fuels. Mines which had been idle were back in operation and other mines had been expanded to produce more. Furthermore, in 1945, the Company started the development of the Greenwood Forty-Foot and Mammoth Stripping, which was the largest stripping operation ever to be undertaken by the Company. One other significant thing happened at the end of 1946. The Edison Anthracite Coal Company, which had leased the Nesquehoning Colliery for five years, terminated its lease in December, 1946.[61] This meant that the coal produced and sold by Edison Anthracite at Nesquehoning now had to be sold by the Old Company's Lehigh sales force. The other Company collieries, for the first time in five years, had to share their sales market with Nesquehoning Colliery. This meant less working time for all collieries. The Equalization Agreement, part of the Union agreement, was now back in play for the Nesquehoning employees.

The high productive capacity of the coal mining properties at the end of the War resulted in fewer days' operation of the mines and strippings to produce the declining market needs. More idle time at the collieries led to increased costs of production; it was estimated that each idle day cost the company $20,000.

Another factor at the end of the War adversely affected the Coal Company: productivity of the miners had decreased. This may have been due to rehiring former employees of the Company discharged from the armed forces. Many of these employees were not coal cutters or miners, so the ratio of producers to non-producers decreased. In 1946, the Company employed 4,982 men compared with 4,643 men in 1945. Production per day at the mines decreased from 15,016 tons in 1945 to 14,356 tons per day in 1946. Little wonder profits were in a downward trend.

The average earnings for the employees of the Lehigh Navigation Coal Company in 1946 was $3,292. The payroll for the year amounted to $16,399,988. In addition, 901 stripping and rock contractor employees received wages of $2,775,879, an average of $3,080 per employee. Contract miners earned an average of $17.32 per day in 1946 while working on contract and while on company work, or "consideration," $10.56 per day. Wage rates had been increased substantially during the War years because the industry (industry bargaining took place) could not afford to be shut down due to strikes. Now with labor rates at an all-time high, productivity had declined and production costs increased considerably. These higher costs could not be recovered because competitive fuels were now able once again to draw customers away from anthracite. The margin of profit was decreasing rapidly.[62]

As profits continued to deteriorate at the Coal Company, White selected Evan Evans as President of The Lehigh Navigation Coal Company on Dec. 23, 1947. J.B. Warriner was appointed Chairman of the Board.

Evan Evans, President, Lehigh Navigation Coal Company, 1947-1952.

Biographical information for the new president is as follows: Evan "Evie" Evans was a native of Coaldale, Pennsylvania, where his parents had settled as immigrants from Penclawdd, Wales. He was one of 10 children. His father, Evan G. Evans, was inside foreman at No. 8 Mine of Lehigh Navigation Coal Company when Evie started to work there as a water carrier in 1909. After serving as a laborer and machinist helper, he started as a chairman on the surveying corps of the Company.

Evie studied engineering with the International Correspondence School, Scranton, Pennsylvania, while working for the company. He continued to advance, serving successively as District Engineer, Assistant Superintendent, Superintendent, Vice President and General Manager, and on December 23, 1947, he was elected President.[63]

Evans was deeply involved in public and civic affairs. He founded the Panther Valley Recreation Center. He was a Director and President of the

Year	LC&N Co. Consol. Profit	LC&N Co. Dividends	Lehigh Navigation Coal Co. Profits	Lehigh Navigation Coal Co. Tons Produced	Lehigh & New England R.R. Profits	Anthracite Industry Production	Comments
1946	$ 2,701,535	$1.00	$222,683	3,935,403	1,118,538	61,978,710	Edison terminated Nesquehoning Lease
1947	2,650,022	1.00	64,018	4,486,341	1,439,066	58,277,985	
1948	3,495,932	1.00	19,181	4,309,060	1,890,494	59,108,810	L.N.C. Co. 6,200 Employees
1949	2,263,118	.50	(475,998)	3,485,000	NA	44,710,118	
1950	2,626,545	.80	97,654	3,635,118	NA	46,339,255	
1951	1,646,820	.70	(848,267)	3,217,702	NA	42,389,055	
1952	2,001,325	.70	(392,887)	3,453,052	1,830,953	40,067,130	October – Weston-Dodson Merger
1953	467,487	– 0 –	(3,361,170)	1,724,704	NA	30,495,391	
1954	(1,398,484)	– 0 –	May 13, 1954 Properties Closed	550,264	1,051,011	26,612,312	October 6, 1954 Properties Leased

Table III. Results of Lehigh Coal and Navigation Company and Subsidiaries in Post-War Years (1946-1954).

First National Bank, Coaldale, Pennsylvania, and Chairman of the Anthracite Section of the American Institute of Mining and Metallurgical Engineers. Evans was a retiring and unassuming person in spite of his many successes and was affectionately known as "Evie" by almost everyone.

Evie, the "hometown" boy, had the confidence and respect of the employees of the company. After the War years, however, the losses for the company continued to mount, and he embarked on a program to increase productivity and eliminate bad labor practices (particularly the miners quitting early). His efforts met with great resistance from the men and the local U.M.W.A. union, with the miners striking time after time.

Evans' good standing in the community didn't influence the miners to cooperate with him in correcting the company ills, and he was unable to stem the large losses. Because of ill health, he agreed to become Chairman of the Board, and turned the presidency over to Glenn O. Kidd in June, 1953.

Table III gives the financial and tonnage results in the post-War years of 1946 through 1954 for The Lehigh Coal and Navigation Company, Lehigh Navigation Coal Company, and Lehigh and New England Railroad.[64] Production of the Lehigh Navigation Coal Company and the anthracite industry are also listed. Results for the first three years after the War for the parent company were fairly satisfactory and a dividend of $1.00 per share was maintained.

In 1949, the bottom fell out of the market for anthracite. Sales for the industry declined from 54,066,000 tons in 1948 to 41,000,000 tons in 1949, a drop of 24 percent in one year.[65] White stated in the 1949 annual report, "It is our opinion that this decline is not an indication of downward trend." However, this was the turning point: except for a slight improvement in 1950, the market continued its downward trend. By 1953, it had dropped to 27,000,000 tons, exactly half of what it was six years before.[66] This was the problem facing the Coal Company subsidiary when capacity was at an all-time high.

Based on White's belief that anthracite would be able to hold its markets, a major stripping operation had been started in 1945. The subsequent development program required considerable money. The Greenwood Forty-Foot and Mammoth Stripping would eventually be 1,500 feet wide, 700 feet deep, and 3,000 feet in length. Several million dollars were tied up in this operation during its early years. The froth flotation plant under construction in 1945 was put into operation in 1946. This plant recovered fine anthracite from the colliery wash water which had formerly been discharged into the streams. The plant clarified the wash water so that it could be reused for coal preparation.

The research department of the company, under the direction of H.S. Gilbertson, had been successful in producing a lightweight aggregate, called

Mr. Robert Taney, Superintendent of the Lelite Plant, describing Lelite, which is a product made from salt and is used as a lightweight aggregate in building construction. The Old Company was a pioneer in this development.

Lelite, from colliery waste. This aggregate made it possible to produce concrete of one-third less weight. The lightweight concrete was useful in the construction of tall buildings, bridge decks, etc. The management of the Company was pleased to be able to convert waste into a useful product, and believed that this development offered a good opportunity for diversification. It hoped that this new venture would favorably impress the investment community and the stockholders.

Accordingly, the decision was made to build a 125,000 ton per year lightweight aggregate plant near the Tamaqua Colliery. Construction of the

Lelite plant began in 1947 and was completed in the following year. Efforts got under way to promote and sell the new product, which was eventually successfully used in many famous buildings, including the White House, Girl Scout Headquarters in Washington, DC, and the Manufacturers Trust Building and the Guggenheim Museum in New York City. Despite the high quality of Lelite, the concept of a lightweight concrete was difficult to sell to architects and the volume of sales for the high priced product was never sufficient to make the Lelite plant a profitable diversification move.

Modern and more efficient coal-cleaning equipment for the recovery of Nos. 4 and 5 Buckwheat coal was put into use at the three collieries in 1952.

New diesel locomotive of the Lehigh and New England Railroad as part of the modernization program.

The fine coal installations at the Coaldale Colliery provided for flotation equipment similar to that which had been pioneered and installed at the Tamaqua Colliery several years previously to recover fine anthracite from the colliery wash water. Another large expenditure went into 800 new steel mine cars to replace wooden cars. Mechanical mine car cleaning devices were also installed to replace hand methods previously employed. At Lansford Colliery, the refuse hauling system was converted from rail to truck.

Several new mine pump installations had been completed on the new mine levels and additions made to older pump stations where increased capacity was needed. Facilities were improved for coal trucking by long distance over-the-road dealers at both Coaldale and Tamaqua Collieries. A Truck Sales Marketing force was formed to further promote truck sales. These moves showed that for the first time the Coal Company was permitted to throw off the shackles of the incestuous relationship it had developed with the Lehigh and New England Railroad.

A new method of mining was developed which employed power drills for drilling long holes to blast down the coal in order to increase miner productivity. Three new stripping developments were also started in 1951 to maintain a high rate of production from this surface mining source.

In addition to the capital outlays made in Lehigh Navigation Coal Company, considerable expenditures were made in Lehigh and New England Railroad. During the War years, the Railroad purchased new covered hopper cars for cement and open steel hopper cars for coal so that it could handle increased tonnages, and more new cars were purchased after the war. In 1947, the Railroad ordered diesel electric freight units for use principally on its main line service. Twelve of these units were delivered in 1948; 20 were added in 1949 to complete the dieselization program, by which the Railroad expected annual savings of $850,000.

The stripping and modernization program was estimated to cost $8,000,000 according to 1948 studies by company engineers and consulting firms. This included $3,000,000 for stripping operations and various improvements in mining properties in 1946. Additional stripping costs and construction of the Lelite plant in 1947 involved an additional $3,130,000. Because of the general situation at the time and difficulty in getting long-term loans on favorable terms, the company financed this part of the program from working capital and short-term loans. A $3,500,000 loan was made from a group of banks in 1947, with $1,000,000 of it as revolving credit to store coal and $2,500,000 as secured loan. These bank loans were repaid and the Company's working capital position restored during 1949 upon the issuance of new general mortgage and collateral trust 4.5 percent bonds in the amount of $6,000,000 due December 1, 1959. The program of improvements would continue, taking approximately two years to complete.[67]

New method of mining employing long hole drilling and blasting techniques. *Courtesy of George Harvan.*

Funded debt of the company had been reduced to $18,434,000 in 1947 under the plan followed by White over the years, but by the end of 1949, funded debt increased to $26,870,473.[68]

The 1949 annual report stated that the Company had paid out $15,452,000 in cash dividends, purchased over $17,710,000 of new plant and equipment, invested large amounts in strippings, and at the same time

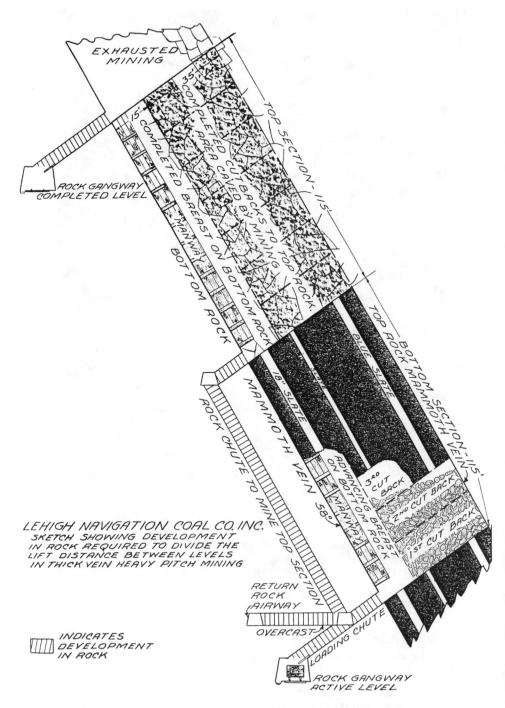

Lehigh Navigation Coal Company, Inc., sketch showing development in rock required to divide the "lift" distance between levels in thick heavy pitch mining.

Tamaqua truck scale. *Courtesy of George Harvan.*

Narrow gauge railroad from Nesquehoning Colliery to Lansford completed in 1948. Railroad made it possible to close the Nesquehoning Breaker.

Miner and his "buddy" drilling in a breast at a mine of the Lehigh Navigation Coal Company. *Courtesy of George Harvan.*

reduced funded debt over the last 10 years. At this stage $5,762,000 was invested in stripping operations.[69]

When the expansion and modernization program was set forth by White after the War years, he had formed a project committee made up of the head of the comptroller's office and the heads of all the parent company's subsidiaries. This committee was charged with the responsibility of evaluating all the projects requiring capital expenditures in all the subsidiary companies. By so doing, it was hoped that the capital funds would be channeled into projects that would yield maximum return. This was a commendable program, but the whole improvement program was based on the false premise that demand for coal would continue at a high rate over the future years.

Problems at the Mines

From 1945 to 1953, White approved major capital expenditures to maintain stripping production, modernize the mining and preparation facilities, and diversify into the production of Lelite, a lightweight aggregate made from coal refuse. Despite the money spent to improve mining operations, results steadily worsened. Table III shows that the Coal Company had a profit of $222,683 the first year after World War II ended. Each year thereafter, except for 1950 when a modest profit was made, results kept getting worse. Staggering losses occurred in 1949, 1951, 1952 and 1953. Why was this trend happening?

The major problem was the excess production capacity at the mining properties while the total demand for anthracite continued to drop. This decline, which had really started in 1927 and continued throughout the depression years up to 1939, was temporarily delayed and actually reversed during World War II years. The impact of war in Europe at that time caused an aberration in the demand trend-line for anthracite that lasted through 1948. After that date the decline continued at a merciless rate. As a result the Lehigh Navigation Coal Company had four large collieries producing great volumes of coal for a rapidly receding market. At the same time the rest of the anthracite industry had been drastically reducing its capacity. In particular, the other major companies such as Glen Alden, Reading Coal, and Hudson Coal Company, had reduced their outputs but not the Lehigh Coal and Navigation Company.[70]

A feeble attempt to consolidate operations was planned for early 1948. When completed this plan eliminated the Nesquehoning Breaker. Coal formerly processed at this plant was hauled by rail from Nesquehoning to the Lansford Colliery. The breaker at Lansford was the newest of the Company's plants and had sufficient capacity to handle the extra tonnage from Nesquehoning.

Employee Problems

As noted earlier, one of the major problems causing high costs at the mines was the poor attitude of the employees. Most of the employees of the Coal Company were descendants of former workers. Sons, grandsons and great-grandsons all worked for the major industry in the area. However, in response to memories of past real or supposed abuses by management, many of the men harbored a deep-seated feeling of hostility against the Company. Each generation inherited a suspicion of management. Irish employees felt they were discriminated against by Welsh and English supervisors; workers of Polish, Czechoslovakian, Hungarian, Russian, and Italian descent also felt a resentment against the Welsh and English bosses. There was also religious and ethnic bigotry.

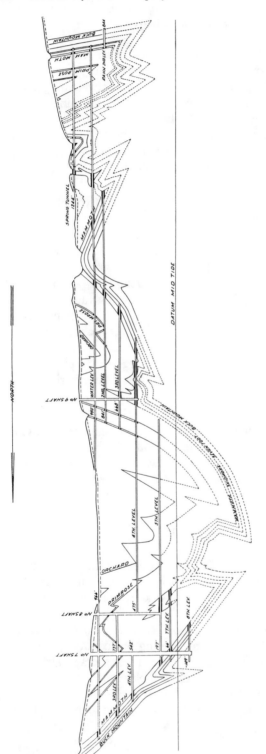

The Lehigh Navigation Coal Company cross section showing veins in Panther Creek Valley between Lansford and Tamaqua.

Murders and the atrocities committed during the period from 1860 to 1870 by the Molly McGuires, a secret Irish society, had been sparked by bad feelings between certain Irish employees and the English or Welsh "bosses." These strong feelings of discrimination were never completely forgotten and can be best exemplified by the story of an Italian worker who was standing in the employment line at the Colliery office seeking a job. Noticing that the only men being hired ahead of him in line had names like Edward Edwards and Thomas Thomas, when his turn came at the employment window and was asked his name, he responded, "Tony Tony."

Jobs in the mines were dangerous, dirty and hard. Even under the best of conditions dust was impossible to completely control. As a result, miners were subject to black-lung or anthrasilicosis. Because of these occupational hazards, many families had relatives who had lost their lives, suffered injuries or had black-lung disease. This strengthened their rejection of mining as a vocation and their dislike for the company that offered employment as miners. One new resident to the mining area observed that the only time he saw the miners smile was when they were on strike.

The accident and death rate among Lehigh Navigation Coal Company employees was consistently below the industry as a whole. Management carried out excellent safety programs and did everything possible to train miners to work safely. Much time and effort also went into attempts to control coal dust. Regardless, because of the nature of mining, accidents did happen and employees were affected by dust and other occupational hazards.

The coal seams in the Panther Valley were steeply pitched as compared with coal seams in the northern anthracite region where they were generally horizontal. Because of the steeply pitching and thick seams, special mining methods had been developed to extract the coal. This method was called the "breast and pillar" method but it actually had many of the characteristics of a "shrinkage slope" in ore mining areas. After drilling and blasting the coal from the face of the breast or cut back, the miners would draw only a portion of the broken coal from the opening created by the mining activity. The remainder of the broken coal was left in the opening to provide a "platform" on which the miners could stand to enable them to drill and blast the next cut. By such a process, they could cut the coal from the solid vein from one level in the mine to the next level. After the "breast" was driven to its limit, it was filled with loose coal and the miners could now draw out the coal from the "breast" by running it down chutes into the mine cars in the gangways or haulageways. Two miners worked together to complete this method. After the miners removed coal from the breast, they would drive chutes into the pillars between the breasts. When the chutes were driven to the top of the pillar, the miners would drill and blast the coal at the top of the pillar so that it could be drawn through a battery into the chute and down to the haulageway where it was loaded into cars. After the coal was removed above, the miners would then drop down the chute some distance, establish a new battery, and once again drill and blast the coal above. This coal could then also be drawn through the new battery into the chute for loading into mine cars. This operation was called pillar "robbing."

Because of the steeply pitching seams, mechanization of the mining operation was not as feasible as in anthracite and bituminous mines which had flat seams of coal. In fact, drilling holes in the coal to insert powder or explosives was done for the most part by using hand drills or augers. Few miners used jackhammers. However, the Panther Valley mines did have the advantage that gravity could be used to load the coal into mine cars. Because of the thick, steeply pitching seams of coal mined by the Company, much of the development work of driving haulageways and chutes was done in the rock formations below the coal seams. This added considerable cost to the production of coal as compared to mines with flat or horizontal seams. There the haulageways could be driven in the coal seam itself, and the coal that was recovered by driving the openings partially paid for the development work.

Two miners worked as a team in each breast or pillar. The foreman in charge of a given section of the mine had to climb up steep chutes or breasts, depending on where the miners were working, to inspect their work and give instructions. Due to the large number of working places under each foreman, the amount of time he could spend with each pair of miners was considerably limited. Accordingly, the miners had to learn to work with only intermittent and limited supervision. The miners developed a great deal of independence.

A payment system that encouraged the miners to work hard with minimum supervision had to be used. For this reason a contract or piece work method was employed. The miners were paid a stipulated amount for each yard of tunnel, chute and breast they drove. Also, they received a fixed amount for each car of coal they produced after the breast "came in" and when they were "robbing" pillars. However, abnormal mining conditions resulted in the development of "consideration work" in 1907 or 1908.[71] Mining sometimes presented situations to which contract rates did not always apply. Originally, foremen dealt with these problems by granting the affected miners a special allowance. This was called "allowance work" later known as "consideration work."

As a result, "consideration work" started as a means of paying a contract miner temporarily on a day rate when the physical conditions of his area were such that ordinary contract rates did not apply. Conditions warranting consideration pay are "loose coal" requiring "fore poleing" to safely drive a chute, or extreme heat, excessive water, excessive dust, etc. Thus, consideration rates were paid originally for a type of work that required exceptional mining skill either because the work was extremely hazardous or because the work was performed under unusual conditions.[72] Initially, such payments were very infrequent.

As time passed, the conditions to which the consideration concessions originally applied were stretched beyond proper limits. Consideration time constituted a major problem because its original purpose was abused. This abuse increased particularly during World War II years when the Company was under great pressure to produce coal at a maximum rate to meet increased demand stimulated by the War effort. Concessions were made by the supervisors to avoid strikes. Accordingly, the men slid into an attitude that they had the right to work at consideration rate instead of contract rate.

Miners had traditionally been permitted to set their own quitting time when they were being paid contract rates for the work accomplished. However, they also attempted to set their own quitting time when they were working on the consideration or day rate. The Brown Warriner Agreement stated that the quitting time for contract miners working under normal conditions on the consideration rate, was 1:30 p.m. for shifts commencing at 7:00 a.m.[73]

The abuse of the contract method of payment and quitting time of the miners caused many destructive and illegal strikes in the Company's Panther Valley mines. The miners expected consideration payments when conditions became the least bit difficult instead of exerting some additional effort to make satisfactory wages on a contract basis. Furthermore, they continued to quit the mines when they chose, while expecting to be paid at the day rate. When they refused to work the time prescribed by the foreman depending on conditions, or until the established 1:30 p.m. quitting time for consideration work, the foreman paid the men either the hours worked at the consideration rate or the contract items, whichever was higher. When the miners were "docked" for not staying the specified time, they would strike. When a strike occurred at one of the four collieries of The Lehigh Navigation Coal Company, a "General Mine Committee," made up of representatives of the various local unions, would meet the same day and vote to support the local union where the grievance had originated. Accordingly, on the following day, all the employees of the Company, including contractor employees, would go on strike.

The abuse of the "consideration" wage was to a great degree the fault of the assistant foremen who were the front line supervisors. Paying miners employed at contract rate in any pay period, for unworked shifts at the consideration rate for the sole purpose of making up wages or justifying early quitting time, was a common practice. By the end of World War II, over 50 percent of all contract miners' payments were at the consideration rate.[74] The contract method of payment was no longer effective in maintaining a high degree of productivity. This was one of the principal reasons for the decline in production at the Company's mining properties. While productivity declined, costs continued to escalate at an alarming rate.

Evan Evans, who was Vice President and General Manager of the Coal Company from 1940 to 1947 and President from 1947 to 1953, was seriously concerned about the drop in the miners' productivity. After World War II, he embarked upon a campaign to correct the consideration payment abuse and the early quitting time problems. As a result, one strike after another occurred at the mining properties.

In July, 1948, the United Mine Workers dispatched an international commission to the Panther Valley to hear the complaint of Coaldale No. 9 Mine workers who were "docked" when they refused to stay in the mines until 1:30 p.m. The commission upheld the action of the Company officials. Lansford No. 6 miners staged a sitdown strike inside the mines and only came out when International U.M.W.A. President John L. Lewis agreed to meet the concerned parties in the Washington headquarters of the union.[75] Once again, the Company position was supported. The Anthracite Board of Conciliation affirmed the 1:30 p.m. quitting time when the No. 6 Mine grievance was referred to them. Many other strikes occurred over this same issue.

The miners continued to ignore not only the orders of their supervisors, but also those of their top international union officers, as well as the findings of the Conciliation Board Umpire. The working time lost through strikes for the Panther Valley mine workers continued to be excessive. This unhappy record indicated that labor conditions at the company mining properties were becoming self destructive. The cumulative effect on the sales force, Old Company dealers, and the consuming public was devastating.

One other problem developed which had an adverse effect on the effectiveness of the contract method of payment of the company miners. All the wage contracts from 1945 to 1952 provided for per diem increases rather than an increase in the contract rates themselves. By 1952, contract workers were paid $6.117 per day in addition to their basic contract earnings, which were calculated on the basis of units of work multiplied by the contract rates in effect in 1945.[76] This gave the miners little incentive to work on a contract basis and induced them to opt for consideration rates whenever possible. This problem was the fault of both Union and Company negotiators: they were virtually destroying the contract system of payment for Panther Valley miners.

Productivity of the hourly employees required to service the needs of the coal cutters or producers and the productivity of the hourly employees working in and about the breakers or preparation plants on the surface were also low.

Probably another reason for low productivity at the mining properties could be traced back to the conflict of interests between the Coal Company and the Lehigh and New England Railroad. Coal Company officials and supervisors felt that their efforts to correct bad labor practices were not being supported by the "Big Office" in Philadelphia. Consequently, supervisory morale was low and many poor practices and conditions were permitted to proliferate over the years. Overall efficiency suffered and costs continued to be high.

The Disastrous Years (1949-1953)

The winter of 1948-49 turned out to be the warmest in 78 years in the Northeastern part of the United States. Because of a decline in coal production and shipments, The Lehigh Coal & Navigation Company showed a consolidated loss for the first half of 1949. The Board decided to postpone consideration of a dividend at that time. By September 30, 1949, the nine month figures showed a fair profit and on October 27, 1949, the Board declared a 50 cent dividend for the year, half of what it had been in recent past.[77]

Since White had become president of The Lehigh Coal & Navigation Company, he had followed a policy of reducing the funded debt. Now a new issue of general mortgage and collateral trust 4½ percent bonds in the amount of $6,000,000, due December 1, 1959, was sold to a group of institutional investors. The funded debt rose to $26,870,473. The increased funds were necessary to pay for the stripping development program started in 1947 and the modernization program, estimated to cost $8,000,000. These programs, described earlier, continued for the next two years.

In 1949, the year White had to raise money by going to the financial institutions to sell the new bonds, was also the year that the total production of anthracite hit a new low of 41,000,000 tons. White explained once again that the large Coal Company losses of $475,998 were due to the "unusually warm winter." Only 3,280,000 tons had been sold.[78]

The Board of Managers of the parent company was growing more concerned with results. The dividend was cut in half and the issuance of additional bonds approved, thus increasing funded debt. The Board, however, went along with White's wishes and recommendations. The Board members in most cases had been recommended for their positions by White, and owed to him their allegiance. At this time, the number of stockholders was more than 13,000, none of whom owned large blocks of shares. As a result, management had not yet felt any substantial pressure from them. The Board was still friendly to White. Doubtless, however, there were more questions in the Board room about large expenditures at the mines and modernization of the Lehigh & New England Railroad. Not too long a time was to elapse, however, before White was to be challenged as large blocks of stock passed into the hands of a few major stockholders. The "White Board of Managers" was soon to come under fire.

In 1950, White had a slight "breather" as production of anthracite picked up slightly with a colder winter. The Board was able to pay 80 cents in dividends since the consolidated profits were $2,626,545.[79] The Coal Company showed a $97,654 profit, and Lehigh & New England Railroad was able to show favorable results because the modernization program enabled it to reduce operating costs.

White had appointed Evan Evans president of the Coal Company when J.B. Warriner retired in 1948. Evans was encouraged by the increased sale of anthracite in 1950 and communicated to employees as follows: "Results show up favorably and it looks as though we are on the sunny side of the street. The demand is brisk and we'll keep the Valley busy. If there is a normal cold weather, it should permit the mines to continue to operate at fulltime schedule."[80]

Walter Banta, vice president in charge of sales, retired in 1950 and Glenn Kidd, secretary and assistant treasurer at the time, was selected his successor. Kidd immediately initiated an aggressive sales program to increase coal sales. He utilized many new sales techniques, such as offering liberal credit terms to dealers. His goal was to enhance the Company's share of the total anthracite market. Prizes and premiums were now offered to dealers for improved sales. One of the areas of concentration was in the industrial sizes of anthracite. The Company was recovering greater tonnages of the finer sizes because of the construction of the froth flotation plants and the improvements in the fine coal facilities at all the Company breakers.[81]

The 1950 Annual Report for the company emphasized how the Company was endeavoring to utilize its assets to enter other markets. The land holdings which had been acquired over a hundred years earlier to assure an adequate supply of water for the canals were being put to use to complement the construction and operations of Split Rock Lodge. The Lelite plant was built to produce lightweight aggregate from waste mine materials. To capitalize on the valuable watersheds owned by the Company, efforts were made to promote the use of water from the Pocono Mountains for the City of Philadelphia. None of these efforts were really successful, however, and did nothing to arrest the downward trend of the Company.

The year 1951 was a bad one for Lehigh Coal and Navigation Company. Profits sank to $1,646,820. The Board decided to reduce the dividend to 70 cents per share as compared with 80 cents the previous year.[82] During this year, White hired H. Louis Thompson to be assistant comptroller of the parent company. Thompson had graduated from Lehigh University and had maintained a close contact with its Business School. He quickly recognized the need for a market study for anthracite and arranged for the University to make such a study. When this study was completed, it indicated that the demand for anthracite would continue to decline to 15,000,000 tons by 1965. This study confirmed his own belief, as well as that of certain other members of the management team, including Glenn O. Kidd, who was now serving as Vice President in Charge of Sales for the Coal Company. White would not believe that the "Lehigh University Market Study for Anthracite" was valid, and refused to agree to a proposed plan of consolidation of the mining operations. Thompson and Kidd could not shake his faith in the anthracite industry.[83]

The Lehigh and New England Railroad was still benefiting from the dieselization and modernization program, and, despite a reduced amount of anthracite freight, it produced a profit of $1,830,953 for the year. Remarkably, this railroad had operated at a profit every year since 1913. The officers were anticipating new freight for the line with the construction of the Pennsylvania Power & Light Company's Martin Creek, Pennsylvania, plant which would require 900,000 tons of bituminous coal annually. Five hundred new coal cars were scheduled for delivery during December of 1951 and January of 1952, to accommodate the increased business. One hundred new cement cars were also on order.[84]

In 1951, the demand for heat, due to unseasonably mild temperatures, averaged approximately 12 percent lower than normal. Competitive fuels, such as oil, and bituminous coal were more available. As a result, further reduction in sales and production at the Coal Company caused a loss of $848,267.[85] At the Coal Company, President Evans was valiantly trying to improve the operating efficiency of the mining properties by tightening up on operating costs. He made top level management changes in 1951 in an attempt to get better results: W.J. Parton was made General Manager; Norman Richards was appointed General Superintendent; D.C. Helms, former General Manager, was appointed to a new position, Vice President in charge of Long Range Planning.[86] All efforts to reduce costs and losses were futile, however, as the demand for anthracite continued to shrink. Sales Vice President Kidd was doing an outstanding job of increasing the Coal Company's share of business, but his efforts were also to no avail. The diminishing overall demand for anthracite continued.

About this time, the Internal Revenue Service placed a 1913 value for income tax purposes of $51,000,000 on the Lehigh and Susquehanna Railroad and the Wilkes-Barre Railroad leases with the Central Railroad of New Jersey. This valuation represented $26.44 per share on the 1,929,109 outstanding shares of common stock. The stock in the Company was selling for $8.50 per share and Model, Roland and Stone, an investment firm in New York, was quick to recognize that the stock was very much underpriced. The Lehigh and Susquehanna Lease was only one of the many assets of the Company. The book value of the Company according to the 1951 Annual Report was only $50,525,672, or $26.19 per share. Obviously, if the Lehigh and Susquehanna Lease asset alone had a value of $26.44, the book value of the Company was understated. Leo Model of Model, Roland and Stone, was aware that the Company was negotiating with the Central Railroad of New Jersey for the sale of the Lehigh and Susquehanna Railroad properties for $25,000,000. A payment of $25,000,000 for the Lehigh and Susquehanna Railroad could provide a tax free liquidating dividend to the shareholders of approximately $13 per share. Model and his firm started to buy Lehigh Coal and Navigation Company stock.[87]

The aggressive sales efforts of the Sales Department of the Coal Company under Kidd's guidance started to pay off in 1952. Coal Company sales were 7.8 percent ahead of 1951, whereas the industry's sales were 11.8 percent lower than 1951.[88] The new management team was also making some progress on decreasing costs by reducing manpower requirements, increasing productivity and, of course, by the benefits derived from the capital improvement programs. The operating loss was considerably reduced.

Because of reduced Coal Company loss, parent company's profits for 1952 were somewhat improved. Consolidated profit was $2,001,325. A 70 cent dividend was declared, the same as the previous year.[89]

Some changes were taking place, however, in the Board of Managers. Two of the older members of the Board, Bruce D. Smith and E.B. Leisenring, died during the first half of 1952. William M. Hickey was elected by the Board of Managers on September 25, 1952, to fill the vacancy created by the death of Smith. Hickey was President of the United Corporation which held 48,705 shares of the Company's common stock.[90]

In October, 1952, the Lehigh Navigation Coal Company purchased the assets of the Weston Dodson Coal Company of Bethlehem, Pennsylvania, which operated anthracite mines and strippings. It was well established in the industry. Under the merger agreement, the sales department of the Weston Dodson Company would now sell coal for the Lehigh Navigation Coal Company properties, in an attempt to increase the Company's share of the business. The merger agreement provided for the purchase of Weston Dodson by issuing 273,000 shares of Lehigh Coal and Navigation Company, the surviving corporation of Weston Dodson.[91] This company now became the largest stockholder.

C. Millard Dodson, president of Silverbrook, a coal mining subsidiary of the Weston Dodson Company, was elected by the Board of Managers on December 22, 1952, to fill the vacancy created by the death of E.B. Leisenring.[92]

Thus, it can be seen that the stock of the Company, for the first time, was now being concentrated in the hands of a few investors. Large blocks of stock were represented on the board by Hickey and Dodson. Furthermore, Model and his firm continued to purchase the Company's stock. The winds of change began to blow stronger.

The year 1953 had to be an agonizing one for White. Certain people on his management team questioned his policy of refusing to curtail the mining operations of the Coal Company despite the reduction in coal sales. His program of investing new capital in the mining properties was also debated. White still believed that the market for coal would come back and industrial sales for anthracite could be increased. In addition, the Board of Managers, always under his influence and control until now, became more doubtful as the year rolled on.

Dodson and Hickey formed an alliance shortly after their election to the Board and were joined by Frederick C. Dumaine, Jr., a member of the

Board since 1948. These men alternately challenged White's management policies; every move was now questioned. White, sensing that his hold on the Board was beginning to slip, tried to convince friendly board members that his policies were the best for the long run. The dissident faction was attempting to get the remaining Board members to side with them. The fight for control was heating up.

The anthracite industry also had a bad year in 1953. Commercial production of all sizes was about 27,000,000 tons, 26 percent less than in 1952. The decline was attributed to expanding competition of domestic and imported fuel oil and natural gas, and to the sixth successive abnormally warm winter. The year 1953 established a new record for the warmest calendar year in the 82-year history of the weather bureau.[93] The lower total production for the industry and the higher temperatures meant more problems for the Company.

White decided that he had to take drastic steps to cut the alarming and growing losses of the Lehigh Navigation Coal Company. On June 13, 1953, he appointed Glenn O. Kidd, Vice President in charge of sales for the Coal Company, to the presidency of that Company. Evan Evans, who was in failing health, was made Chairman of the Board.[94] W.J. Parton, John C. Bolinger, and Richard C. Newbold were appointed Vice Presidents. Newbold was put in charge of sales, filling the job vacated by Kidd.

Kidd immediately started to work on drastic cost cuts at the mining properties. One of his first steps was to fire or pension 65 salaried personnel and institute a cutback in salaries of all department heads. The Lansford shops were consolidated, eliminating 150 employees. Selling expenses were also reduced. On November 4, 1953, Kidd arranged for a meeting of officers of United Mine Workers of America, including Martin Brennan, the Union's district president.[95]

At the meeting with Union officials, Kidd discussed the tremendous 1953 losses of the Coal Company. He stressed that the rapidly declining market demand for anthracite resulted in excess production facilities of all coal companies in the industry. Intense competition developed out of this, causing very low selling prices for coal. Non-union mines (bootleg mines) were able to produce and sell coal at $5 per ton less than the legitimate companies. Kidd told the Union representatives that production dropped from over 54,000,000 tons in 1948 to 27,000,000 tons in 1953. Employment in the industry was down from 88,000 in 1941 to 51,000, but the Lehigh Navigation and Coal Company had virtually no reduction in force over that same period of time. The company could not survive without shrinking its operations in the face of the drastic loss of market for the entire industry.

Kidd then unveiled plans for curtailing losses. Nesquehoning Mine and Dodson Breaker, formerly Lansford Breaker, were to be closed at the end of the current heating season. The Lansford mines and strippings would have to be closed if conditions worsened, but if they did not, the production from these mines and strippings would be transported to Coaldale for processing. Secondly, an immediate drive was to be started to increase the daily rate of deep mine production by 20 percent. A written statement was placed in the hands of 40 U.M.W.A. representatives and later in the hands of 160 foremen. A tape recording of the proceedings was to be broadcast over Radio Station WLSH, Lansford. The meeting with the Union officials was a stormy one since 650 men would lose their jobs under Kidd's program. However, the Union agreed to cooperate.[96]

Lehigh Navigation Coal Company officers in June 1953: Glenn O. Kidd, President, Evan Evans, Chairman of the Board; and Richard Newbold, Vice President of Sales.

Despite all the efforts to minimize losses in 1953, industry sales declined 26 percent and prices were badly depressed. Excess production capacity of the industry and the Company was the cause. A record loss for the Coal Company of $3,361,170 occurred.[97] Production for the Lehigh Navigation Coal Company dropped to a new low of 1,724,704 tons. Idle time costs of $20,000 per day were causing disastrous results.

The tremendous losses undermined the support for White at his Board of Managers. Profits from the railroads and other subsidiaries of the parent company looked as though they would be wiped out by the Coal Company losses.

By year end, the consolidated net income for Lehigh Coal and Navigation Company was $467,487 (including a tax carryback credit and related adjustments of $334,514). The Board of Managers decided to pay no dividends for the year.[98] And so a period of 72 years of unbroken dividend payments was shattered.

Mr. Scattergood, another older member of the board, died in mid 1953, and the Board was forced to discuss his replacement. White rejected a candidate who had been proposed by Model. Robert Dechert, partner of the prestigious law firm of Barnes, Dechert, Price, Myers and Rhoads, in Philadelphia, was recommended by Dodson and his backers on the Board. White knew that Dechert was a well thought of, reputable lawyer in the Quaker City, and finally acceded to the wishes of the dissident board members. White believed that "maybe, Dechert will be able to talk some sense into them." Attorney Dechert was elected to the Board of Managers on November 25, 1953.[99]

But White was wrong. Shortly after Dechert was elected, he asked to meet White. At this meeting, he said he was representing the majority of the Board in asking White to resign his position as President. White first refused, but later agreed in order to avoid being voted out of office at the next Board meeting. He decided to negotiate the best deal possible for himself rather than fight with the rest of the Board.

As part of the maneuvering to determine who would head the Lehigh Coal and Navigation Company, Glenn O. Kidd, who was now serving as President of the Lehigh Navigation Coal Company and had been R.V. White's "fairhaired boy" and understudy to succeed him, visited Millard Dodson. Kidd attempted to convince Dodson that he should be the new President of the parent company and that Dodson should take over the presidency of the Coal Company subsidiary. Kidd argued that he had been groomed for the top job whereas Dodson was particularly well qualified to operate the coal properties because of his previous experience with the Weston Dodson Company. Dodson didn't buy Kidd's suggestion and responded with an unequivocal "no."

Under the arrangement worked out with the Board of Managers, White agreed to the appointment of C. Millard Dodson as Executive Vice President on January 28, 1954. White also agreed not to stand for reelection as President of the Company at the organization meeting of the Board of Managers following the stockholders' meeting in April, 1954.[100] He negotiated a contract with the Board under which he would serve as Chairman as long as the Board desired him to do so; and would render consulting and advisory services to the parent company or any of its subsidiaries on request of the new President or the Board of Managers. This contract was subsequently ratified at the following stockholders' meeting.

On December 5, 1953, Marshall S. Morgan died. The vacancy on the Board created by his death was filled by Leo Model. Model and his investment firm had been buying the Company's stock since 1951 or 1952 and now held in their own name or in their clients' names approximately 20 percent of all the outstanding shares. Model thus represented the largest single block of stock in the Company, and his influence on future events of the Company would be considerable, to say the least.

The takeover of the management of the Lehigh Coal and Navigation Company was actually consummated when Dodson was appointed Executive Vice President in January, 1954. He and his allies on the Board of Managers had complete control from that point. The most disappointed person in the company was Glenn O. Kidd who, up to this time, had fully expected to succeed R.V. White.

Before leaving the "White Years," I shall discuss in greater detail certain factors relating to White's fall from the presidency of Lehigh Coal and Navigation Company in 1953 as expressed to the author by H. Louis Thompson.[101]

1.   The Lehigh Coal and Navigation Company was negotiating with the Central Railroad of New Jersey for the sale of the Lehigh and Susquehanna Railroad properties for $25,000,000. The negotiations had proceeded to the stage where the price was agreed upon, but stalled when the Central Railroad of New Jersey was unable to obtain financing. White and his management team were disappointed when the Central Railroad of New Jersey was unable to come up with the necessary money to consummate the deal, and negotiations were finally called off.

2.   The Lehigh Navigation Coal Company operating losses were increasing at an alarming rate in the last few years. These losses were only partially offset by earnings of the Lehigh and New England Railroad.

3.   White refused to accept the fact that many anthracite markets were permanently lost and that demand would continue to shrink. His misguided optimism prevented the Company from establishing a strategic retrenchment plan based on the shrinking total demand for anthracite. If such a plan had been instituted early enough in his tenure in office, the large expenditures for capital improvements and increased capacity could have been channeled into making the most productive operations more efficient while at the same time phasing out the higher cost operations. With such a strategic plan, the Company could have shifted some of its resources into diversified fields and reduced its dependence on the coal production transportation complex which made up its main business.

4.   The Public Utilities Commission was expected to hand down a decision permitting the trucking of cement from the Lehigh Valley cement plants. Projections had been made that when this occurred, the Lehigh and New England Railroad would lose 50 percent of the freight business from the cement mills. This was borne out at a later date. The reduction in the cement hauling business would make the Lehigh and New England Railroad operate at a marginal profit even if the anthracite tonnage remained constant, but anthracite tonnage continued to decline, and the Lehigh and New

England Railroad would operate at a loss. Since assets of $22,000,000 were employed by the railroad, this situation would not make economic sense.

5.   A group in Wall Street had analyzed the picture at the Lehigh Coal and Navigation Company and came to the conclusion that the Central Railroad of New Jersey deal would be eventually consummated for $25,000,000, and that the Lehigh and New England Railroad could also be liquidated for its book value of approximately $22,000,000. Because of the potentially large liquidating dividend from such a strategy, the interested group in Wall Street pursued a very active stock buying program.

This group teamed up with the Dodson interests and were responsible for White's loss of control.

THE DODSON YEARS (1954-1959)

C. Millard Dodson, President, Lehigh Coal and Navigation Company, 1954-1959; Manager, 1953-1959.

C. Millard Dodson and the other directors representing large blocks of Lehigh Coal and Navigation Company stock were now in complete control of the Company. Dodson and several generations of Dodsons before him had been identified with the coal industry for many years. He previously spent 15 years as President of the Weston Dodson Company, a family-owned business founded in 1837 at Bethlehem, Pennsylvania. Dodson had attended Yale University and Babson Institute, and served as a Naval aviator from 1942 to 1945 with the rank of lieutenant commander.[102] He was blunt and to the point, did not mince words, and was willing to confront problems headon. One of his long-time employees described him as having "ice water in his veins." Such was the man who took over the reins of much troubled Lehigh Coal and Navigation Company.

Changes at the Lehigh Navigation Coal Company

Just as the management of the parent company was changed suddenly at the end of 1953 and early 1954, so also was the management team of the Lehigh Navigation Coal Company just as quickly altered. Dodson didn't think that Glenn O. Kidd and he could work together: He asked for Kidd's resignation. Kidd had just been reelected president of the Coal Company at the Coal Company's Board reorganization meeting on March 23, 1954. On April 30, 1954, he and one of his vice presidents, John C. Bolinger, Jr., were gone. Dodson appointed W. Julian Parton to the office of President and Joseph J. Crane as Vice President and General Manager. Parton and Crane were given the formidable task of formulating a plan to eliminate the overwhelming losses incurred by the mining properties, particularly in 1953. The program then had to be sold to the U.M.W.A. officers and to Company employees. On behalf of the Board of Managers of the Lehigh Coal and

Navigation Company, Dodson ordered Parton to suspend for an indefinite period all mining operations. Operation of the Coal Company was to be resumed only after a satisfactory plan of operation was approved by management and U.M.W.A. officials.[103]

Parton, the mining engineer who had joined the Coal Company in 1939 directly from graduate school, had guided the development of a new method of mining, long hole drilling and blasting, which offered great hope for increasing mine productivity. He had substantially improved results at the Nesquehoning Colliery during his brief tenure as Superintendent before he was promoted to Assistant General Manager. Could this young, 38-year-old mining engineer bring about the changes in work habits and attitudes that had been allowed to deteriorate for the 134 years of the Company's existence?

Parton's first move was to notify District President Brennan of the U.M.W.A. that mining operations of the Lehigh Coal & Navigation Company would be suspended for an indefinite period because of operating losses. Following this prenotification, the shutdown of the collieries was announced on May 3, 1954. On May 7, President Parton and associates met with Brennan and other U.M.W.A. aides to give their views of the parent Company's attitude toward Coal Company operations. They discussed two possible plans by which the parent company might be persuaded to reopen the mines. Brennan asked Parton to meet with the full membership of the Panther Valley General Mine Committee on May 8 to apprise them of the situation. On the day of the meeting with District U.M.W.A. officials, Parton also met with the Company department heads, colliery superintendents and the general mine foreman to explain the situation.[104]

On May 8, 1954, management met with officers and committee members of all Panther Valley local unions, together with the district leadership, at the Old Company Club in Lansford. Parton explained the two plans which he and his staff had prepared for operation of the collieries without sustaining losses. If one of these plans was approved by Union officials, in order for resumed operations to be successful, he would need assurances of cooperation to obtain increased productivity and other operating efficiencies. During the stormy meeting, representatives of the local unions castigated the Coal Company management and the parent company for previous practices.

In a later, private meeting of the U.M.W.A., Brennan pledged his cooperation in helping the Company management get the mines back to work and they decided to refer the Company report to the individual local unions for action. On this same day, a comprehensive statement was issued by the Company for publication in local newspapers, explaining the reasons for the shutdown and the need for assurances from the Union.

On May 10 and May 11, all local unions held special meetings, widely publicized as sessions to decide whether to "accept or reject" the Com-

Lehigh Navigation Coal Company President W. Julian Parton and Vice President Joseph Crane studying Coaldale Breaker in connection with Plan 2 to consolidate operations and thus reduce losses of the coal company, in spring of 1953.

pany's so-called Plan 2 for reopening the mines. Nos. 10 and 11 locals took decisive action by accepting the Plan 2 "under protest." The other locals balked and voted to send telegrams asking John L. Lewis to come to the Valley. On May 11, a meeting of the Panther Valley General Mine Committee was convened at Lansford, and a canvas of local union meeting results showed no clearcut acceptance of the Company's proposal. The Committee did dispatch a wire in the name of the General Mine Committee to John L. Lewis, requesting him to come to the Valley. This telegram was subsequently acknowledged by U.M.W.A. Vice President Thomas Kennedy, a Lansford native, who stated that Lewis had referred the matter to him. While stating he was currently tied up in Washington, he promised he would meet with the committee as soon as possible.

President Parton attended an executive committee meeting of the parent company management in Philadelphia to report that up to this point he had not been assured by the U.M.W.A. that full cooperation would be forthcoming. As a result, the committee had taken no action on reopening the mines. This meeting and its result were widely publicized in Valley newspapers.

Parton issued a notice to department heads on Friday, May 14, in which he furloughed all monthly personnel effective the next day. Four days later, C.M. Dodson, at a gathering of 50 community leaders assembled at the Mahoning Valley Country Club, presented a candid report on the financial situation of the Company and cited the uncertain outlook for future mining operations in Panther Valley. He recommended strongly that the community leaders concentrate efforts to bring new industry into the Valley since the mining industry would not be able to sustain the economic health of the area any longer.

Parton's Plan to Operate the Mines

The detail of Plan 2 was as follows: All mines, including Nesquehoning, were to reopen. All strippings would suspend operations temporarily. Tamaqua Breaker was to be converted to a "headhouse" on a single shift to enable loading of material from Nos. 10 and 14 mines into gondolas. Coal from No. 11 mine was to be shipped directly to Coaldale Breaker. One half of the Lansford head house was to be used to load material from Nos. 4 and 6 mines into gondolas. Headhouse material from both Lansford and Tamaqua was to be shipped to Coaldale Colliery. Coaldale Breaker was to operate two shifts to prepare all mine coal, including that from Nos. 8 and 9 mines.

Parton's plan to operate deep mines in lieu of strippings was contrary to the current trend in the entire anthracite industry. However, it would employ the most workers and, in Parton's judgment, it was "the proper thing to do."

Parton established the following conditions to be accomplished if the plan was to succeed: Miners would have to produce as much coal as they were capable of producing each day by working longer hours than in the past. Consideration payments would be eliminated except under genuinely abnormal conditions. Improved mining methods such as mechanization and recently developed long hole drilling and blasting must be expanded as quickly as possible. Working places, where practical, were to be double shifted to improve the maximum productive employment. The agreement between the Lehigh Navigation Coal Company and the U.M.W.A. was to be strictly adhered to in all details. Illegal strikes must not take place. Mine committees were to limit themselves to the functions authorized by the general contract and agreement. Interference by mine committee members beyond their rights under the general contract and agreement had to cease.

Thus it can be seen that Parton's Plan 2 involved producing all the coal mined in the Panther Valley through one breaker at Coaldale, which was located in the center of the Valley. Also, Parton was banking on increased productivity in the mines by eliminating many bad practices. The basic contract method of paying the miners was to be restored. Consideration or day rates were to be paid only when conditions were abnormal and the miners couldn't possibly make satisfactory contract wages.[105]

International U.M.W.A. officers involved in Panther Valley labor problems in 1953: Thomas Kennedy, Vice President; Santo Volpe, Union official; John L. Lewis, President; and Martin Brennan, President, District 7, U.M.W.A.

U.M.W.A. Vice President Thomas Kennedy met with members of the Panther Valley General Mine Committee on May 20 in the Hazleton offices of the Union. No authorized information on the results of the meeting was revealed to the Company or the public. The following day Parton issued a brief statement emphasizing the fact that the Company manager, after three weeks, still had no official word from the U.M.W.A. regarding the pledge of cooperation and increased productivity, and the other questions which the company had raised.

Kennedy and the General Mine Committee arranged to meet with officials of the Company on Friday, May 28, at the Old Company's Club. The situation to date was reviewed: Kennedy literally "pulled the rug" from under his district aide, Brennan, by restating the U.M.W.A. position as seeking to restore the mining operations to the status held prior to the May 3 shutdown. He asked for a meeting with C.M. Dodson and the parent company Board of Managers for the purpose of telling them that the plan to curtail strippings and rock work as proposed by Parton was wrong. A letter requesting such a meeting was to be sent to Dodson.

After the May 28 meeting, Parton and his associates conferred with Dodson and his aides. They decided to notify District President Brennan that the Company intended to announce on May 29 the schedule of mining operations under so called Plan 2, with the exception of Tamaqua. Dodson attempted to call Kennedy directly at Hazleton, but was unable to reach him.

On May 29, Brennan told Parton that he and Kennedy had recommended that Plan 2 be put into operation and that a General Mine Committee meeting had been called that (Saturday) evening. On the same day, a full page announcement was made in the Valley newspapers incorporating Parton's statement that mining operations would restart on Tuesday, June 1. An accompanying statement was published over the name of Dodson. News stories prominently featured highlights of these announcements. There seemed to be general approval and acceptance of the "good news."

The Panther Valley General Mine Committee met that Saturday evening and recommended that the local unions vote upon the plan announced earlier in the day. Midway through the meeting, Brennan called President Parton and requested that the start of work be deferred until June 3, to give all local unions a full opportunity to meet. The Company agreed to this schedule.

On June 1, the local unions met. Again, no decisive majority voted in favor of the return to work. Tamaqua Local voted to wire President Lewis asking him to visit the Valley and made known its intentions to stop the work which had been started two days earlier to convert the breaker to a head house. Coaldale Local voted to return to work "under the same conditions as existed when work was stopped" and also issued telegrams to Lewis and Kennedy asking them to come to the Valley to negotiate with the Company. Coaldale No. 8 Local also sent an additional wire to Kennedy asking whether any of the so-called five points laid out by President Parton violated the contract. On the same day at 4:00 p.m., the Panther Valley General Mine Committee voted against a return to work under the five-point plan and sent another wire to U.M.W.A. President Lewis asking him to attend a mass meeting at Coaldale High School stadium "at his earliest convenience."

On the night of June 1, Coaldale No. 8 Local had its second meeting of the same day. This time the Local reversed its stand of the previous night after hearing a telegram sent by Kennedy which advised that the Coal Company management's proposal, in his opinion, "does not conflict with the General Agreement" and said that he "urged the resumption of operations to the extent provided," i.e., under the so called Plan 2. After a heated argument, the Local voted (57 to 32) to comply with Kennedy's advice, and the Local's mine committee was instructed to report its action to the General Mine Committee. On that same day, Dodson received a letter from Thomas Kennedy requesting a meeting with the parent company management.[106]

The Panther Valley General Mine Committee, in response to its telegram to John L. Lewis asking him to attend a mass meeting in Coaldale, received a telegram from Lewis tersely informing the committee that "mass meetings are a thing of the past" and ordering an immediate resumption of work in Panther Valley.[107] On the following day, June 3, the Panther Valley General Mine Committee called a special session to hear the contents of the Lewis telegram. Once again, this Committee referred the matter back to all the local unions, asking them to meet between June 3 and June 5.

The local unions met on June 4 and agreed to return to work by a five to one majority, with Tamaqua No. 14 Local casting the negative vote. Coaldale No. 8 agreed to return to work with the same reservations as previously, "work under the 1952 contract."[108]

A return to work looked encouraging after the Panther Valley General Mine Committee met and heard the reports of the various local unions. The General Mine Committee agreed to resume work on Monday, June 7, 1954.

On Sunday, June 6, rumors of picketing were widespread throughout the Valley. Accordingly, no one was particularly surprised when a motorcade of pickets, made up of Tamaqua Local members and led by officers and mine committeemen of the Local, prevented resumption of work at Coaldale and Lansford collieries. General Mine Committee officials were reported to be "noticeably disturbed" at the development and complained to the newspapers that "nobody had told them that they (Tamaqua) were going to picket us on Monday."

On the following day, June 8, pickets continued to prevent work at the mines. A mass meeting was called, probably encouraged by No. 14 Local, for 2:00 p.m. at the Lansford High School stadium. The meeting was well attended and conducted in an orderly fashion. Mine committeemen and local union members of Tamaqua No. 14 Local were the main speakers. Charges of appeasement and conniving were leveled at District President Brennan, and there was implied criticism of John L. Lewis. The speakers condemned Parton's plan for reopening the mines and charged that the company "locked out" the men of Panther Valley and intended to take away many conditions previously enjoyed under the contract. Several speakers referred to subversive elements and "refugee money" used for buying and controlling many shares of Lehigh Coal & Navigation Company stock, and called for a probe of the parent company management by Congress.[109]

A seven-man committee was appointed by Tamaqua No. 14 Local to meet John L. Lewis on June 10. On that date, newspapers said an investigation by the Justice Department of Lehigh Coal and Navigation Company's stock transactions had been requested by U.S. Representative Francis E. Walter, whose district included Carbon County. Walter said he had written a letter to Attorney General Herbert Brownell. Simultaneously, U.S. Representative Ivan D. Fenton, whose district included Schuylkill County, announced he had requested the Securities and Exchange Commission to investigate charges made by Coaldale No. 8 Local that funds of a refugee group were used in acquiring and controlling stock of Lehigh Coal & Navigation Company with the intent of liquidating the company holdings and realizing a "fast dollar" on the deal. Also on June 10, Tamaqua No. 14 Local received a letter from John L. Lewis calling the picketing earlier in the week "a serious mistake and certainly brought about rule by the minority instead of the majority."[110] The Local's request for a special meeting with him was rejected by Lewis, who said he would "be glad" to meet only after operations resumed and "your local union and its members withdraw its action of opposition to resumption."[111]

Lewis did meet with representatives of seven Panther Valley local unions in Washington on June 14, and he asked again that work resume. He spoke to the group for an hour, but did not grant Tamaqua Local representatives a special audience as they had requested.

At a special meeting of Tamaqua No. 14 Local the next day, the representatives who had attended the Washington meeting on the previous day charged Lewis with giving the Tamaqua committee "evasive answers" concerning protests that Lehigh Coal & Navigation Company's plan to restart the mines was unjust. The committee also reported that Lewis told them "every operator is doing the same thing."[112] The local rejected the Lewis back to work order and voted to continue picketing the Coaldale and Lansford collieries.

Lewis actually attended a Tamaqua local union meeting at a downtown lodge hall and was highly incensed when the rebellious local membership shouted down his appeal for reason. The imperious Lewis was not accustomed to this disrespect. Angry, he left the Valley that night and was reported to have said he was washing his hands of this impossible band of troublemakers.

The Tamaqua Local once again met in a special session on June 16, and voted to ask other anthracite U.M.W.A. districts to join in its fight and to hold a mass meeting at the Tamaqua High School stadium on Sunday, June 20.

Mass meeting of Tamaqua Colliery miners at Tamaqua High School stadium—one of several such meetings in 1954. *Courtesy of George Harvan.*

On June 17, C.M. Dodson went to Washington to meet personally with John L. Lewis to review the details of the Company's financial situation. Lewis indicated he planned certain steps in an effort to clear up the Panther Valley situation. The Securities and Exchange Commission announced on the same day that the Lehigh Coal and Navigation Company's stock dealings had been given a clean bill of health. Coaldale No. 8 Local also published a statement that it "has not agreed to work under Julian Parton's five point program, but that it has agreed to return to work under the terms and conditions of the 1952 agreement."[113] On the day after Dodson's trip to Washington to meet with John L. Lewis, the fiery U.M.W.A. leader met with members of the Panther Valley Mine Committee in Hazleton. He again asked for work to resume and for pickets to be withdrawn and gave short shrift to the Tamaqua No. 14 Local's protests.[114]

On Saturday night, June 19, Tamaqua No. 14 Local held another special session which was attended by Martin Brennan, President of District 7, and David (Red) Stevens, a Lansford resident and international officer, representing Lewis. The Tamaqua Local received a report that the Company had told Lewis that mining operations would be closed permanently on Monday, June 21, if the miners did not return to work by then. Lewis had requested a postponement of this announcement. Lewis' representatives, Brennan and Stevens, were subjected to boos and catcalls at this meeting.[115]

Approximately 1,000 people attended the Tamaqua Local's mass meeting at Tamaqua High School stadium on Sunday, June 20. The principal speaker was the editor of *The Coaldale Observer*, ex-Congressman James H. Gildea, who had played such an important part in setting up the Work Equalization movement in the early 1930's. Gildea castigated the Company's management and called for a committee to confer with Pennsylvania Governor John S. Fine and President Dwight D. Eisenhower. He also recommended a region-wide tieup and talked about "militant pickets" who would prevent any removal of machinery or equipment from the mines. The idea of naming a committee to confer with Fine and Eisenhower was approved.[116]

W.J. Parton spoke over the radio to the people of the Panther Creek Valley on Monday, June 21. He restated the pertinent details of the plan by which he had hoped to put the mines back to work on May 29. He emphasized that the central question of restarting the mines on a profitable basis had been sidetracked largely by misrepresentations and accusations. He stated that "unless something transpires between the time of his broadcast and Thursday, June 24, he will have to report to the Board of Managers at Philadelphia that he has failed."[117]

A Citizens' Committee comprised of Tamaqua No. 14 Local union officials, Editor Gildea, Carbon County Judge James C. McCready, and Carbon County Commissioner Ambrose O'Donnell met with Governor Fine in Harrisburg on June 23. After a two-hour session, the Governor called Dodson at Bethlehem and relayed the Committee's request to put off any curtailment of mining operations until the Committee could hold a mass meeting in Panther Valley on Sunday, June 27, at 2:00 p.m. Dodson thanked the Governor for his interest and consideration and told him that all the facts would have to be presented to the Board of Managers at Thursday's meeting.[118]

The Tamaqua Local scheduled a meeting for Thursday night, presumably to receive a report on the Harrisburg conference with the Governor. It also announced plans for the mass meeting on Sunday at 2:00 p.m. at the Lansford High School stadium.

Decision to Discontinue Lehigh Navigation Coal Company Mining Operations

The Company's Board of Managers met in Philadelphia on Thursday, June 24, and decided to discontinue mining operations of the subsidiary firm, Lehigh Navigation Coal Company. Thus the chapter on Company operation of the Panther Valley mines for 134 years was finally closed.[119]

Tamaqua No. 14 Local's special meeting on the same day that the Company announced the Board of Managers' action to cease operating the mines largely ignored the closing order. The Local instead voted approval of a 10-point program for operating the mines as conceived by Carbon County Presiding Judge James C. McCready. Floyd Eveland, a mine committeeman in Tamaqua, is quoted as saying, "The company is bluffing, trying to put fear in the hearts of men." The Local also proceeded with plans for the mass meeting on the following Sunday.[120]

However, the Company wasn't bluffing. In anticipation of possible trouble at the Panther Valley operations when the announcement that the parent company was suspending mining in the Valley, Dodson asked Parton and his family to leave the Valley for a week. They left, but the precaution wasn't necessary: throughout the entire period of May 3 to June 24, when meeting after meeting was held and tempers were strained as the pros and cons were argued, no acts of violence occurred in the Valley. This is a real credit to the former employees of the Lehigh Navigation Coal Company. Unfortunately, they were too steeped in the tradition of unionism to consider crossing the picket lines of the one dissident local union. The majority wanted to recognize the facts as they all too plainly showed, but resisted.

With the decision made to suspend mining operations by the Lehigh Navigation Coal Company, President Parton was given the job of administering the mining properties. The properties were to be maintained in operating condition for the time being. This meant that the dewatering pumps in the mines would continue to operate to prevent workings from flooding. Parton and a skeleton force stayed on to study what could best be done with the mining properties. Selling the mining facilities or leasing them seemed to be the only avenues open for salvaging something from the investments the Company had in these properties.

The people of the Panther Creek Valley were subdued for the next few months; nothing seemed to be happening. Former employees were trying to get used to the fact that the Lehigh Coal & Navigation Company had actually done what had so often been threatened over the long history of the Company. Many of them undoubtedly thought that the Company was still bluffing.

In the latter part of August, 1954, Parton had a visitor, Joseph (Buck) Tout, who had been chairman of the mine committee at Nesquehoning when Parton was appointed superintendent at Nesquehoning Colliery in 1947. Parton and Tout had a lasting friendship and mutual respect for each other. Tout said that he was representing the men at both Nesquehoning Mine and Lansford Colliery in asking Parton to organize a company to lease the Lansford Colliery, including the Nesquehoning Mine. The men promised to do whatever was required of them to get back to work. Parton saw this contact as an opportunity to get things moving again in the Valley. He discussed this proposal with his former Vice President, Joseph Crane, and James and Frank Fauzio, stripping contractors at that colliery. They agreed to organize a company and to lease the Lansford and Nesquehoning operations from the Lehigh Coal and Navigation Company.[121]

Panther Valley Coal Company Lease of Lansford Colliery

About the time the operations of the mining subsidiary ceased, another subsidiary, the Lehigh Navigation Coal Sales Company, staffed by key personnel from the sales department of the coal producing firm, was formed for the specific purpose of assisting prospective lessees with their sales problems.

Because of the promise of full cooperation from the Lansford and Nesquehoning colliery employees and the opportunity to have a sales outlet for any coal production, the four founders of the Panther Valley Coal Company took an option to lease the Lansford Colliery and Nesquehoning Mines. If this option were exercised, an exclusive sales contract was to be negotiated with the new Lehigh Navigation Coal Sales Company.

James Fauzio, Joseph Crane, W. Julian Parton, Frank Fauzio, Panther Valley Coal Company officials who resumed mining operations in October, 1953, by leasing Lansford and Nesquehoning Collieries.

As soon as the option was granted, the officers of the Panther Valley Coal Company negotiated a new contract with the U.M.W.A. International U.M.W.A. Vice President Kennedy insisted on negotiating this contract with the new Company. For the most part, most of the previous practices that were so detrimental to successful operation of the mines in the Valley were eliminated in this new contract.[122]

On September 8, 1954, the Panther Valley Coal Company exercised its option to lease the Lansford Colliery and Nesquehoning Mines.[123] Work was immediately started to make necessary repairs to the breaker and mines to put this colliery back into operation. On October 6, 1954, production was resumed.

Everything looked encouraging for the new Company. The miners were producing at a considerably greater rate than before, and costs were favorable. The sales volume was adequate to keep the Lansford Colliery operating on a full schedule since the sales company now marketed the output from one colliery; previously it had sold the output from three collieries. Furthermore, truck bins had been installed at Lansford breaker to handle sales of coal to the trucking trade. In the past bins had been installed only at Coaldale and Tamaqua.

Coaldale Mining Company Lease of Coaldale Colliery

The euphoria didn't last too long. James H. Pierce, who headed a consulting firm in Scranton, Pennsylvania, organized the Coaldale Mining Company and leased the Coaldale Colliery from Lehigh Coal & Navigation Company in November, 1954. Pierce had a longstanding relationship with John L. Lewis. It was widely reported that he had financial backing of the U.M.W.A. through the Riggs National Bank in Washington, D.C. of $2,500,000 and eventually up to $8,000,000.[124]

The resumption of production from the Coaldale Colliery by the end of the year immediately had the effect of cutting into the sales of the Panther Valley Coal Company. Truck sales dropped off and, once again, more production was available than the market for anthracite could fully absorb.

Dodson, however, was pleased that he was able to state in his first L.C.& N. annual report in 1954 to the stockholders that two leases had been made before year end "bringing to 70 percent the total of properties formerly worked by us now under lease."[125]

At the beginning of 1954, the Company owned six active retail coal yards which were adding to operating losses at the rate of approximately $80,000 annually. These unprofitable properties were disposed of in line with a management policy of eliminating losses.

Because of the downward trend in anthracite, the new management of Lehigh Coal & Navigation Company concluded that the rate of depreciation taken on the books for the anthracite mining properties should have been greater. Accordingly, the Board decided that a reserve was desirable for the revaluation of the anthracite properties. This reserve, as reflected on the balance sheet for 1954, was $10,500,000, for total revaluation of mining properties and for the Lelite plant.[126]

The financial statements at the end of 1954 indicated a consolidated net loss of $1,398,484 was incurred over the year.[127] With the revaluation of the mining properties as shown in Table IV, the assets decreased to

Evan Thomas, Sales Manager, Lehigh Navigation Coal Company, inspecting the first car of coal from the Panther Valley Coal Company, October 6, 1954. *Clarence Dankel.*

$67,740,254 from $86,903,904. The net worth of the Company decreased to $40,620,645 from $54,327,901 in 1953. The longterm debt was $21,254,464, a decrease of $1,000,000 compared with the previous year. There was no dividend in 1954.[128] The Lehigh Navigation Coal Company also showed a huge loss of $5,290,689 in 1954. Total production for the year, including production from leased properties, was only 826,146 tons.[129]

President Dodson told the stockholders in his annual report that "there is much evidence to support the feeling that we have trimmed sail and set a course which will permit the company in 1955 to get out from

Happy Nesquehoning miners returning to the "pits" after the reopening of the mines under lease to the Panther Valley Coal Company on October 6, 1954. *Clarence Dankel.*

under a steady burden of operating losses." He promised, "if such proves to be the result, your Board of Managers will, of course, give earnest consideration to the resumption of dividend payments within the limits of prudent financial policy and debt service requirements."[130]

So ended Dodson's first year as president of the Lehigh Coal and Navigation Company. He had made unprecedented changes to eliminate the heavy losses at the anthracite mining properties. Would these drastic steps be successful in resolving the financial problems of the Lehigh Coal and Navigation Company?

Miners handing in lamps to Joseph Gover, lampman at Nesquehoning Mine at end of first day mines reopened on October 6, 1954. *Clarence Dankel.*

Dodson's second year as President of the Lehigh Coal and Navigation Company was not quite as turbulent as the first year. Having gotten rid of the burden of operating the coal properties by leasing 70 percent of the Company's coal lands to independent operators, the Company was able to show a profit by the end of 1955 and to resume dividend payments. [131]

However, changes were still in the making at the coal mining properties. The Panther Valley Coal Company was starting to feel the effects of competition from the Coaldale Mining Company which had leased the Coaldale Colliery. The truck business of the Panther Valley Coal Company was the first to be affected because Coaldale Mining Company had a more favorable location and more efficient truck facilities. [132]

The main problem, of course, for both lessees of the coal properties was the continuing downward trend in the anthracite industry. Production for the year again showed a decrease; unstable prices continued to prevail. An increased tonnage of coal was being mined by nonunion sources. This low-cost production had an adverse effect on the profits of the operations. Competitive fuels, especially natural gas, made deeper inroads.

As sales continued to lag and prices softened, the Panther Valley Coal Company started to show losses for the first time since it was organized. It was unable to continue to meet its obligation to pay 50 cents per ton into the Health and Welfare Fund of the U.M.W.A. [133]

THE LEHIGH COAL AND NAVIGATION COMPANY AND SUBSIDIARY COMPANIES

CONSOLIDATED BALANCE SHEETS, December 31, 1954 and 1953

ASSETS:	1954	1953
CURRENT ASSETS:		
Cash on hand and demand deposits in banks	$ 4,977,349	$ 2,629,127
U. S. Government obligations, at cost	$ 582,154	$ 929,410
Notes and accounts receivable, including notes of $2,489,591 at December 31, 1953, received under advance sales contracts for deliveries of coal after that date	$ 4,160,135	$ 8,355,033
Less:		
Notes receivable discounted	211,787	1,544,181
Reserves for accounts doubtful of collection	443,648	330,646
	655,435	1,874,827
	$ 3,504,700	$ 6,480,206
Inventories:		
Coal, at the lower of cost or market	297,664	2,780,731
Materials and supplies, at cost	1,253,043	1,877,364
	$ 1,550,707	$ 4,658,095
Refundable Federal income taxes due to carry-back loss		929,034
Other current assets	$ 1,852	$ 46,041
Total current assets	$10,616,762	$15,671,913
INVESTMENTS ($19,020,481 in stocks and bonds of subsidiary companies, eliminated in consolidation, are pledged under funded debt):		
Pledged under workmen's compensation or funded debt	520,191	510,442
Other	10,547	34,763
	530,738	545,205
FIXED ASSETS:		
Coal lands, mining and marketing properties (leased in part in 1954), and Lelite plant (Note 1)	29,368,800	30,003,840
Railroad property (leased in part; leases pledged under funded debt)	49,418,158	49,407,422
Lehigh Canal and equipment	1,548,393	1,545,030
Water property (reservoirs, pumping stations, distribution systems, etc.)	2,560,956	2,605,431
Real estate and equipment, Split Rock development	1,629,553	1,583,722
Other real estate and equipment, not used in operations	879,126	1,004,475
	$85,404,986	$86,149,920
Less reserves for:		
Depreciation and amortization	22,628,681	21,947,612
Depletion, since December 1, 1937	1,286,610	1,256,629
Revaluation of mining properties and Lelite plant (Note 1)	10,500,000	
	$34,415,291	$23,204,241
	$50,989,695	$62,945,679
PREPAID EXPENSES AND DEFERRED CHARGES:		
Prepaid stripping expenses (Note 1)	6,495,623	8,596,519
Less special write-down equal to reduction of Federal income taxes arising from additional tax deductions related to stripping development as permitted under the Revenue Act of 1951	1,621,241	1,621,241
	4,874,382	6,975,278
Unamortized debt discount and expense	249,757	314,565
Other prepaid expenses and deferred charges	478,920	451,264
	$ 5,603,059	$ 7,741,107
	$67,740,254	$86,903,904

The accompanying notes are an integral part of these statements.

THE LEHIGH COAL AND NAVIGATION COMPANY AND SUBSIDIARY COMPANIES

CONSOLIDATED BALANCE SHEETS, December 31, 1954 and 1953

LIABILITIES:	1954	1953
CURRENT LIABILITIES:		
Notes payable, bank	$ 140,601	$ 2,285,245
Accounts payable	2,399,024	2,677,108
Coal deliveries to be made on advance sales contracts	137,257	823,263
Accrued salaries and wages	385,951	680,661
Accrued and withheld taxes	434,381	
Funded debt, mortgages, contracts payable and payments to sinking funds due within one year	1,648,288	1,726,107
Other accrued and current liabilities	776,539	943,607
Total current liabilities	$ 5,922,041	$ 9,135,991
LONG-TERM DEBT:		
Funded debt outstanding	22,287,464	23,922,052
Less held in treasury	1,033,000	181,000
In hands of public	21,254,464	23,741,052
Real estate mortgages and contracts payable	623,829	848,089
	21,878,293	24,589,141
Less obligations and payments to sinking funds due within one year, included in current liabilities	1,648,288	1,726,107
	$20,230,005	$22,863,034
DEFERRED LIABILITIES:		
Workmen's compensation and occupational disease claims determined	303,376	255,840
Other deferred credits	50,167	43,996
	$ 353,543	$ 299,836
RESERVES FOR:		
Workmen's compensation and occupational disease	291,665	258,373
Loss on disposition of marketing assets	315,441	
Contingencies		12,000
	$ 607,106	$ 270,373
MINORITY INTERESTS	$ 6,914	$ 6,769
CAPITAL:		
Capital stock, $10 par value:		
Authorized 3,600,000 shares		
Outstanding 2,202,127 shares	22,021,270	22,021,270
Capital surplus, as annexed (Note 1)	18,599,375	13,916,122
Earnings retained in the business, as annexed (Notes 1 and 2)	None	18,390,509
	$40,620,645	$54,327,901
	$67,740,254	$86,903,904

The accompanying notes are an integral part of these statements.

Table IV. The Lehigh Coal and Navigation Company and Subsidiary Companies' Consolidated Balance Sheets, December 31, 1954 and 1953.

From the time James Pierce leased the Coaldale Colliery, he had been pressuring Dodson to have the Lehigh Navigation Coal Sales Company sell some of the coal produced at Coaldale. Since the Lehigh Navigation Coal Sales Company had entered into an exclusive sales agreement to market the output of the Panther Valley Coal Company in order to encourage this Company to resume mining operations in the Panther Valley, Dodson was unable to comply with Pierce's request. Pierce decided to take matters into his own hands and asked for an appointment with Parton. At this meeting, Pierce demanded that the Panther Valley Coal Company share its sales outlet or he was going to raise the old equalization issue in the Valley. To gain his objective, he threatened more picketing and labor problems.[134]

Merger of Panther Valley Coal Co. and Coaldale Mining Co.

Since the Panther Valley Coal Company was poorly financed and had heavy debt as a result of the purchase of diesel locomotives needed to haul Nesquehoning mine and stripping coal to the Lansford breaker, this Company was hardly in a position to bargain. Parton discussed Pierce's threat with Joseph (Bucky) Tout, his contact with the Panther Valley Coal Company employees. Tout, in turn, contacted the union representatives of the Nesquehoning, Lansford Nos. 4 and 6 mines, and was assured that the Panther Valley Coal Company employees would resist any effort of the Coaldale Mining Company employees to shut them down by picketing. Nevertheless after Parton conferred with his associates in the Panther Valley Coal Company he decided that he did not want to be responsible for any more labor disturbance in the Valley. Accordingly, he contacted Pierce and offered to sell the Panther Valley Coal Company stock to the Coaldale Mining Company for the exact amount invested by the founders of the Company when it was organized. This offer was accepted, and in May, 1955, the merger of the Coaldale Mining Company and the Panther Valley Coal Company was completed. James Pierce, longtime confidante of John L. Lewis, was now in complete control of the Lansford and Coaldale Collieries.[135]

As a result of this merger, W. Julian Parton resigned as President of Panther Valley Coal Company and decided to leave the Valley. He was deeply saddened that he was unable to carry on the company he had organized to get mining activities moving once again in the Valley. However, he had the satisfaction of knowing that his efforts were responsible for the continued operation of two of the three former collieries of the Lehigh Navigation Coal Company.

Parton was retained as a Consultant by the Lehigh Coal & Navigation Company to study a German process for beneficiating fine anthracite and forming the anthracite into a "cokelike" material by the blending of the anthracite with heavy oil before coking the blended material. This investigation involved a trip to Peru where the "anthracite coke" was being produced at Chimbote. In October, 1955, Parton accepted a position as assistant to the President of The General Crushed Stone Company, in Easton, Pennsylvania. Thus ended his career in the coal mining business.

The Lansford and Coaldale collieries of the former operating subsidiary, Lehigh Navigation Coal Company, were being operated under lease, but the third, the Tamaqua Colliery, was still idle. This property was kept available in case another operator wanted to reopen all or part of it. This involved a cost which included power costs for continuing to pump water from the mines, necessary maintenance, and real estate taxes. Negotiations were being conducted with county and local tax assessment authorities for the purpose of having this tax burden reduced.

Railroad income continued to be good. Lehigh and New England Railroad Company made a net profit of $1,952,426. Central Railroad of New Jersey paid the annual rent of approximately $2,300,000 for its use of the Lehigh and Susquehanna Railroad system in 1955. In this year, the Lehigh and New England Railroad handled 2,328,174 tons of cement, approximately the same as in 1954 when an all-time high was established. In 1956, as a result of the expansion of five cement mills in the Lehigh Valley, the Company anticipated an 11 percent increase in cement shipments and a corresponding increase in inbound shipments of raw material. The railroad had forecast a five percent increase in bituminous coal tonnage in 1956 due to the growing needs of the cement mills and the expected completion of a second unit of Pennsylvania Power and Light Company's power plant at Martins Creek, Pennsylvania.[136]

In anticipation of the additional cement business, 300 new box cars were ordered in July, 1955. Additional covered hopper cars for handling bulk cement were also provided by a mutual exchange agreement with a southern railroad. The railroad continued to improve its facilities by installing a new type of scale and by acquiring the latest mechanical devices for tie renewal. Maintenance of way work was thus mechanized as far as possible. Despite wage and material cost increases, lower freight rates on cement effective December 1, 1955, and decreased shipments due to Hurricane Diane, the railroad subsidiary continued to anticipate a prosperous future.[137]

At the annual meeting in 1955, the stockholders approved the creation of a $10,500,000 reserve fund on the Company's books for revaluation of the Company's coal mining properties. The net book value of the mining properties, after the establishment of this reserve, was $3,900,000. Management planned to take larger amounts than heretofore for depreciation, depletion and amortization of deep mine development in accordance with Federal income tax laws and regulations. This meant that, for tax purposes, the remaining value of the mining properties of $14,600,000 as of December 31, 1954, was to be written off in the years 1955 through 1964.

Consolidated net earnings after taxes in the next 10-year period would be increased by the amount of tax reduction which resulted from the larger deductions for depreciation, depletion and amortization of deep mine development. At the end of this period, the anthracite properties would have been fully written off for tax and book purposes.[138]

Since this situation could have great effect on taxability of future dividends, the management requested a detailed analysis of the parent company's earnings and profits by its certified public accountants. This highly technical analysis was completed late in 1955. It showed as of December 31, 1954, no earnings and profits had accumulated since March 1, 1913. The 1955 annual report contained the following statement: "Provided the parent company alone does not have any accumulated earnings and profits for tax purposes in any future years, it is the opinion of counsel that under the current Federal Income Tax laws such portions of distributions by the parent company to its stockholders as may exceed the profits and earnings of the parent company alone for the calendar year in which the distribution is made will not be taxable as ordinary income to the recipient."[139]

The company made a consolidated net profit of $2,422,486 in 1955 as compared with a loss of $1,398,474 in 1954. This led to the resumption of dividend payments after a lapse of two years. On December 27, 1955, a dividend of 60 cents per share was paid to stockholders. This dividend was, in the opinion of counsel and the company's certified public accountant, not taxable as ordinary income.[140]

President Dodson stated in his annual report that two main problems which confronted the company in the past two years had been solved. These problems were: (1) heavy coal mining losses and (2) clarification and study of the various aspects of the Federal income tax position of the company. Dodson stated, "With these problems behind us, your management is now actively studying the possibility of new growth. One avenue is the future development of present assets and another is future diversification. We recognize the advantages of both and are looking for opportunities which may prove of interest. At the present time, there are no definite developments which we can pinpoint."[141] Dodson's statement explained his philosophy of management which apparently was not in full concert with certain other major stockholders whose main objective was liquidation of the company's assets. These diverging views of where the company would go were headed for a collision course in the years ahead and resulted in still more changes in top management of the company.

Expansion and Diversification under Dodson

President Dodson considered 1957 the year in which the Lehigh Coal and Navigation Company had "set sail" for a new course. In his annual report, Dodson stated, "During 1957 substantial progress in our program of expansion and diversification was achieved." The Boone County Coal Corporation, which owned and operated underground and surface bituminous mines in Boone and Logan Counties, West Virginia, was purchased in September, 1957, for $4,073,500. This acquisition consisted of a 23,000-acre area of coal-bearing lands estimated to contain 182,600,000 tons of recoverable coal, a modern preparation plant, underground mechanical mining and transportation machinery, surface mining equipment and necessary above-ground facilities and buildings to conduct an efficient operation. Daily productive capacity of washed coal was 5,500 tons. Production for calendar year 1957 was 1,300,000 tons.[142]

Dodson happily reported that during the short time period involved since the company acquired the property, it had produced results better than anticipated. Another result not reported was the breach that developed over this acquisition between Dodson and Leo Model, the two largest stockholders of the company. Model was opposed to its purchase, and this break between the two would widen in the next few years.

Another subsidiary company, the Penn Navigation Company, completed the purchase of transoceanic shipping interests. The basic assets of Penn Navigation were five dry-cargo "Liberty" ships, a T2 class oil tanker and a 32,650 d.w.t. tanker under contract for construction. The overall cost of the new tanker, including financing charges and other expenses, was approximately $11,000,000, of which 87½ percent was expected to be financed by a mortgage on the ship.[143]

Referring to the acquisitions, Dodson said, "These are the first tangible steps in the company's plan to assume new and important responsibilities in serving industry both from a production and a transportation point of view on a national and international basis instead of concentrating on the furnishing of anthracite coal for domestic heating in the Eastern United States."[144]

After many unsuccessful attempts to sell the valuable Lehigh and Susquehanna properties, the Company executed an agreement with the Central Railroad Company of Pennsylvania (a subsidiary of the Central Railroad of New Jersey) for the sale of the Wilkes-Barre and Scranton and the Nesquehoning Railroads, and the acquisition by the latter of the Susquehanna and the Tresckow Railroads. All these properties had been operated by the Central Railroad of New Jersey for more than 80 years under a perpetual lease from the Lehigh Coal and Navigation Company.[145]

The proposed transactions involved the payment of $3,510,377 to the Lehigh Coal and Navigation Company at the closing and the retention by the Company of its rights as lessor for a period of 35 years. The annual rent which Central Railroad of New Jersey as lessee would pay to Lehigh was $2,283,482, a slight reduction in the rent paid up to that time. Consummation of the transaction was subject to the necessary corporate approvals and to rulings by the Treasury Department and the Interstate Commerce Commission. A special meeting of the stockholders was to be convened at the appropriate time to act upon the transaction. The agreement called for the closing of this transaction no later than June 30, 1958.[146]

Was it possible that the long-awaited goal to liquidate the Lehigh and Susquehanna Railroad properties would be accomplished? Everything looked encouraging.

The Lehigh Coal & Navigation Company entered into an agreement with the Philadelphia National Bank for a loan of $3,200,000 to pay off $1,184,626 of the remaining general mortgage and collateral trust bonds due December 1, 1959, and to use $2,000,000 to complete the purchase of the assets of Boone County Coal Corporation.[147]

Lehigh and New England Railroad results for the year were adversely affected by a month-long strike in the cement industry and the loss of rail shipments of anthracite. Nevertheless it realized a net income of $1,956,466 for the year. During 1957 the railroad's sales force was expanded by the addition of a representative in St. Louis and the appointment of an industrial development agent in Bethlehem. The railroad thus attempted to increase "overhead" tonnage by locating more offline offices and to encourage new industry to locate along the line of the railroad.[148]

On December 31, 1957, the Panther Valley Coal Company advised Lehigh Coal & Navigation Company that it was terminating the lease on the lands which it had been operating. Apparently its U.M.W.A. "shadow" supporters decided that the losses were too great for this company and that it could not continue its financial support. The Coaldale Mining Company, Inc., also reportedly financed by the U.M.W.A., was to continue operations and was expected to produce all the anthracite which could be sold from the company properties in 1958.[149]

Consolidated net income of Lehigh Coal & Navigation Company after taxes in 1957 was only $2,715,425 compared with $3,474,558 in 1956. Part of this decline in earnings after taxes was due to the fact that no loss carryforward was available in 1957 as it had been in 1956. Regardless, the Board of Managers voted a dividend of 75 cents a share. The 1957 dividend was declared not taxable as ordinary income to the recipient.[150]

Undoubtedly the big question in Dodson's mind at the time was whether or not his efforts at diversification would pay off. Anthracite sales for the industry were continuing to decline. New sources of income were greatly needed to offset the Lehigh and New England Railroad's loss of revenue. Another matter of concern was the sale of the Lehigh and

Susquehanna Railroad properties. Would the current agreement be approved by the Treasury Department and the Interstate Commerce Commission? Would Dodson be able to keep the company's largest shareholder, Leo Model, satisfied with the policy of diversification he was pursuing?

According to Dodson, results of the Lehigh Coal & Navigation Company in 1958 were adversely affected by the general business recession which had started late in 1957. Consolidated net income was only $1,683,539, off by more than a million dollars from the previous year.[151]

Dodson discovered that reviving the fortunes of the company wasn't going to be a simple task. The long sought after sale of the Lehigh and Susquehanna Railroad properties did not take place by the stipulated closing date of June 30, 1958, since the rulings of the Treasury Department and the I.C.C. had not been received. Even though the Central Railroad of Pennsylvania was no longer bound by the terms of the agreement, the Company continued to press for a favorable ruling from the Internal Revenue Service with the aim of ultimately consummating the sale.[152] This was a great disappointment. Anthracite business was also off by 16 percent in 1958 and bituminous demand was down 19 per cent throughout the industry. Export tonnage, on which the newly acquired Boone County Coal Corporation leaned heavily, plummeted from 58,000,000 tons in 1957 to 38,000,000 tons in 1958. As usual, the decreased demand caused prices to deteriorate. The bituminous operation had a loss for the year. The only bright spot was the Lehigh and New England Railroad which managed a $1,575,197 profit despite a considerable decline in operating revenues. Operating economies were credited with making this possible.[153]

The annual report made no mention of how the newly acquired transoceanic shipping subsidiary, Penn Navigation Company, fared in 1958. One can only assume that this venture did not do well. Otherwise, some reference to it would surely have been made.

According to H. Louis Thompson, who became President of the Lehigh Coal & Navigation Company in 1960, the rift between the two largest stockholders of the Company, Dodson and Model, had widened with the failure of the Company's new acquisitions to do well. In particular, the purchase of the Boone County Coal Company had become the main cause of discord between the two men. Model had opposed this acquisition, but Dodson insisted on going through with it. Eventually each contact between the two men ended in acrimony. H. Louis Thompson found that he was serving more and more as an intermediary: a good friend of Dodson and on friendly terms with Model, he found his position intolerable. Things got so bad that Thompson's health was clearly affected.[154]

Developments in the Company were not good. Dodson's efforts to straighten out the Company had not been successful, and more serious problems were in the making as 1959 approached.

The Lehigh Coal and Navigation Company took a drubbing once again in 1959: the downward trend in anthracite continued; lost freight revenue affected the railroad subsidiary's income. But another serious development occurred that year which was to have a serious impact on the Lehigh and New England Railroad. The Pennsylvania Public Utilities Commission had in previous years prohibited truck deliveries of cement from the Lehigh Valley cement plants. In 1959, the P.U.C. reversed its former rulings, and truck deliveries of cement started shortly thereafter.

Until this time, the Lehigh and New England Railroad had enjoyed a virtual monopoly by shipping almost of the all output from the Lehigh Valley cement mills. This was a fatal blow to the railroad subsidiary which in its heyday proudly maintained a very high percentage of its most profitable business, "originating freight tonnage", on its line. Most of this tonnage was anthracite and cement, and neither of these would ever again be shipped over the Lehigh and New England Railroad in the same volume as previously.[155]

Anthracite industry production fell from 20,976,000 tons in 1958 to 19,319,000 tons in 1959, a drop of about eight percent.[156] The principal lessee of the Company's anthracite properties, Coaldale Mining Company, Inc., terminated its lease on February 29, 1960. Losses of this company reportedly had been absorbed by the U.M.W.A. The losses were so excessive that they taxed the Union's treasury, and the Lewis regime had "to call it quits."[157] The bituminous industry, still hurt by the loss of the export market in 1958, was also affected by the steel strike. As a result, the newly acquired Boone County Coal Company showed another loss for the year, although somewhat less than in 1958.[158] Obviously, the Company's involvement in both anthracite and bituminous coal was in trouble.

The P.U.C. ruling allowing truck haulage from cement plants in the Lehigh Valley motivated management to try to sell the Lehigh and New England Railroad, as well as the Lehigh and Susquehanna Railroad properties. Accordingly, negotiations underway with the Central Railroad Company of Pennsylvania, a subsidiary of the Jersey Central Lines, were expanded to include all of the Company's railroad interests. No progress was reported, but Dodson promised to pursue "every approach to the problem of its railroad interests in order that definitive action may be taken."

With all the problems encountered in 1959, the consolidated net income of the Company fell to $1,477,580, compared with $1,638,539 for 1958. In spite of the decline, Dodson and the Board of Managers approved a cash dividend of $1.00 whereas the previous year no dividend had been paid. This dividend was deemed not taxable as ordinary income for Federal income tax purposes.

During the year, the fight between Dodson and Model intensified. Dodson's diversification moves, opposed by Model, didn't work out. This gave Model the opportunity to assume an "I-told-you-so" attitude, but relations were so strained between Dodson and Model that all effective communication between them on Company problems completely disintegrated. Thompson was primarily interested in the Company's problems, but also found himself in the position of trying to reconcile their opposing views.

Thompson's health finally gave out and he took a leave of absence. Several weeks later, he received a call from Hans Frank, a member of the law firm representing Leo Model. Frank asked Thompson and his wife to meet with him to discuss Thompson's point of view about the situation, as well as his assessment of the differences of opinion between Model and Dodson. Frank told Thompson that Model wanted to make certain that he (Thompson) would remain with the Company since Model had decided that Dodson "had to go." Thompson was asked to prepare the form of employment agreement he would need to stay on with Lehigh Coal & Navigation Company. Thompson did prepare, in his words, a "very good agreement" which he presented to Model, who accepted it without reservation.

The new understanding clearly eliminated Thompson from the taxing role of acting as mediator between the two principal stockholders. With this understanding and agreement, he was able to resume his duties with the company. Furthermore, the agreement between Model and Thompson set the stage for Dodson's eventual resignation from the presidency of the company. Such was the picture as the company entered 1960.

The president's letter to the stockholders in the 1960 annual report contained the following significant statement: "Recent years have brought changes calling for a reappraisal of your company's affairs with respect to its founding concept as a coal mining and coal transportation enterprise when organized nearly a century and a half ago. Thus it was that 1960 saw attention devoted to the disposal of certain assets that in the view of your Board of Managers did not justify continuance measured against the test of probable long term earnings."

This meant that the Company had agreed to dispose of the anthracite properties in the Panther Valley to a new lessee, the Greenwood Stripping Corporation owned by the Fauzio family of Nesquehoning, Pennsylvania. This agreement was designed ultimately to return to the Company $4,000,000, the net carrying value of the anthracite properties. (See Epilogue for a brief history of the Fauzio family.)

The Lehigh and New England Railroad, the company's principal source of operating income over the years, was put out of business as a subsidiary of the Company when an agreement was reached in June, 1960, selling certain trackage and equipment to a subsidiary of the Central Railroad of New Jersey. A petition was filed with the Interstate Commerce

Commission which requested approval of the sale and permission to abandon the balance of trackage not sold to Central. When approved, the Lehigh and New England Railroad went out of the railroad business. The net liquidating value of the railroad was expected to be approximately $21,045,602, the book value as of December 31, 1960.

. Strangely enough, the liquidation of the Lehigh and New England Railroad took place before the sale of the Lehigh and Susquehanna Railroad properties. Efforts to dispose of these leased properties had been attempted for years, with no success.

Even with the progress made in liquidating the assets of the company by Dodson in the early part of 1960, relations between Dodson and Model didn't improve. If anything, they worsened when Model pressured Dodson to dispose of the Boone County Coal Company. Dodson refused to do this, and eventually Model decided that it was time to remove Dodson as President.

Model asked Thompson to come to his New York office. Thompson was told that Model would have Dodson discharged at the next Board of Managers' meeting unless an arrangement was able to be worked out with Dodson before then. Thompson was asked to relay the message and ultimatum to Dodson. When Dodson and Thompson were in New York shortly there after, Dodson did agree to resign and gave Thompson his written resignation with instructions not to return it. Dodson's two principal backers on the Board of Managers, Hickey and Dumaine, were also in New York at the same time and tried to talk Dodson out of going through with the resignation. Dodson, however, stuck with his earlier decision.

Thompson visited Model at his office and told him that Dodson had agreed to resign. The resignation was accepted by the Board of Managers in September, 1960.[159] F.C. Dumaine, Jr., and William M. Hickey resigned in October and November, respectively. Thompson, Executive Vice President and Treasurer, was elected President, succeeding Dodson.[160] The road was now clear for complete liquidation of the Company's assets as originally contemplated and planned by Model.

No time was lost in disposing of the Boone County Coal Company bituminous property. Partial disposition was effected in December, 1960, when all of the machinery, equipment, materials and supplies were sold. At the same time, the coal lands were leased and an option was granted to the principals of the lessee corporation to buy the coal lands in their entirety.[161]

By the end of 1960, the Company still had a 55 percent interest in the Penn Navigation Company. Because of the depressed condition of the shipping industry, the carrying value of the shipping business investment was reduced by $8,000,000 to bring the figure into line with realistic values.[162]

The company, through several of its subsidiaries, maintained large holdings of real estate in the Pennsylvania Poconos. Real estate activity in these mountains was high in 1960, and the Company planned to step up its efforts in this field.

Split Rock Lodge on Lake Harmony had a successful year. Winter sports at Split Rock and Big Boulder ski areas enjoyed their greatest peak of popularity. Expansion of the lodge was under way.

Robert V. White, Chairman of the Board from the time he stepped down as President on April 1, 1954, in favor of Dodson, died in June, 1960.[163] With Dodson's resignation shortly thereafter, the two former antagonists left the Lehigh scene.

THE THOMPSON YEARS (1959-1969)

H. Louis Thompson, President, Lehigh Coal and Navigation Company, 1960-1968; Manager, 1954-1970.

Thompson, the man behind the scenes during the latter years of R.V. White's presidency of the company and during all of Millard Dodson's, was now President of the once famous "Old Company." Since Thompson was the choice of the principal shareholder, Leo Model, the direction of the Company was already determined. Thompson had little choice but to liquidate the remaining assets of the Company. Thompson, in his typical frank manner, described himself as a "high-priced corporate undertaker."[164]

A Lehigh University graduate, Thompson held responsible positions in other businesses and had been a special agent with the F.B.I. for five years before joining the Lehigh Coal and Navigation Company as Assistant Comptroller in 1951. An expert in accounting and tax matters, he was responsible for many of the sophisticated financial matters pursued by previous managements. He had been the advisor, philosopher, and confidant of the two previous presidents. Unquestionably, he had the confidence of Leo Model.

Thompson sums up his first full year as President by stating that "the management had been actively engaged in a program of disposing of certain assets that in the opinion of your Board of Managers did not merit retention when measured by the yardstick of longterm earnings potential."[165] A substantial portion of that program was completed in 1961.

The sale of the Lehigh and New England Railroad to a subsidiary of the Central Railroad of New Jersey, agreed to in 1960, was approved by the I.C.C. on October 31, 1961.[166] It was completed on that date, and the Lehigh and New England Railroad ceased operations.

The railroad had a net book value of $20,229,420 as of December 31, 1958. As the property was converted into cash, it was expected that "no less than its net book value" would be realized over a period of three to five years.[167]

The Greenwood Stripping Corporation, lessee of the anthracite coal properties in the Panther Valley, constructed a modern automated breaker on the premises in 1960. This plant was outstandingly successful in 1961 when it produced 780,000 tons of coal.[168] The output from this company was marketed by the Lehigh Navigation-Dodson Company under the "Old Company's Lehigh" trademark.

With termination of the Lehigh and New England Railroad operations, the company basically now had left three wholly owned operating units: Lehigh Navigation-Dodson Company, Split Rock Lodge, and Blue Ridge Real Estate Company.

Split Rock Lodge, a year-round vacation and sports center, had a profit of $149,018 in 1961. The Blue Ridge Real Estate Company held a large percentage of the lands in the Pocono Mountains (40,000 acres) and continued to develop and sell parcels of land from its holdings.[169]

The company had lessor interests in the anthracite properties under lease to the Greenwood Stripping Corporation. Through its wholly owned subsidiary, the Lehigh Boone Land Company, it also had a lessor interest in an extensive bituminous coal property in West Virginia.

The company also had a 55 percent interest in the Penn Navigation Company which operated a shipping business. This company operated at a profit of $771,640 in 1961, largely as a result of the sale of three Liberty ships. During 1960 and 1961, two T-2 tankers were converted into dry cargo ships with the hope that they could be operated successfully in the dry cargo trade.[170]

Despite all previous efforts to dispose of the Lehigh and Susquehanna Railroad properties, these properties were still under lease to the Central Railroad of New Jersey for a rent of approximately $2,300,000 per year.[171]

By year end, the Company had long-term debts of $8,188,880. Assets were $54,437,830 and net worth was $43,129,312. The net profit for the year was $522,037. Despite the reduced profit, stockholders received a cash distribution of 60 cents per share, amounting to $1,321,276 in total payout. Company's counsel stated that 32 cents of this distribution was not taxable as ordinary income.[172]

Thompson concluded his 1961 general report with these words, "Now that your company has significant progress in its conversion program, it can address itself to a realignment of its activities."[173]

The Plan of Reorganization

During the first five months of 1962, Thompson and his staff were busy working out the details of the Plan of Reorganization of the Lehigh Coal and Navigation Company, executed on May 31, 1962. Basically, the plan called for the segregation and transfer of certain assets from the Lehigh

Coal and Navigation Company to a subsidiary company called the L.N.C. Corporation. A net worth of $22,781.07 was transferred to the L.N.C. Corporation in return for 100 percent of the common stock of that corporation.[174] The stock acquired by the Lehigh Coal and Navigation Company was then distributed to its shareholders. At the time of the stock distribution, two separate nonaffiliated corporations were established with identical stockholders. However, in as much as the stock of these two corporations was traded in the open market, the makeup of the stockholders would inevitably change.

Basically, the Plan of Reorganization left the Lehigh Coal and Navigation Company only one substantial asset: its investment in the Lehigh and Susquehanna Railroad system, in addition to some other property to be liquidated.[175] All the other potentially active assets had been transferred to the new L.N.C. Corporation.

The consolidated balance sheet, May 31, 1962, for the Lehigh Coal and Navigation Company, Table V, shows that the remaining assets of this company were only $20,004,515, of which $18,951,537 was the book value of the Lehigh and Susquehanna Railroad system.[176]

Stockholders received a cash distribution of 25 cents per share at the time of the reorganization.

The New Lehigh Coal and Navigation Co.

For the stripped-down Lehigh Coal and Navigation Company the fiscal year ending May 31, 1963, was a decisive one according to Thompson. On May 9, 1963, the sale of the Lehigh and Susquehanna Railroad property to the Reading Company was finally completed. Under this arrangement, the Company retained a 35-year leasehold interest in the railroad properties, with an annual rental of $2,283,482 to be paid to it.[177]

The stockholders were informed by letter on May 23, 1963, that the stock of the Lehigh Coal and Navigation Company was delisted from the New York Stock Exchange. It remained, however, on the Philadelphia-Baltimore-Washington Stock Exchange and would be traded both on the Exchange and in the Over-the-Counter market. Since the stock was selling considerably below book value, the management asked for and obtained authorization to purchase stock up to a total of $1,500,000 within one year from the date of the meeting. Management planned to continue purchasing stock with stockholder approval over the period of time from 1963 through 1967. Eventually 317,688 shares were to be acquired at a total cost of $2,123,001 or an average cost of $6.68. This stock was held in the Company's treasury at cost.[178]

Thompson mentioned in his previous letter to stockholders that management was investigating possibilities of replacing the earning power of

THE LEHIGH COAL AND NAVIGATION COMPANY

BALANCE SHEETS

May 31, 1963 and May 31, 1962

ASSETS:	1963	1962
CURRENT ASSETS:		
Cash	$ 393,004	$ 650,000
U.S. Government securities, at cost	1,140,877	—
Notes and accounts receivable	103,503	7,500
Other current assets	—	22,460
Total current assets	$ 1,637,384	$ 679,960
Noncurrent receivables, from sales of assets	$ 460,000	—
INVESTMENTS:		
In wholly owned subsidiaries (Note 1)	—	2,897,344
In 50 pct. owned affiliate	225,000	225,000
	$ 225,000	$ 3,122,344
Leasehold interest, Lehigh and Susquehanna Railroad system (Note 2)	17,335,678	—
Less reserve for amortization	30,624	—
	$17,305,054	—
FIXED ASSETS, AT COST OR LESS:		
Lehigh and Susquehanna Railroad system (leased) (Note 2)	—	$16,056,707
Canal property, etc.	85,738	142,658
Less reserve for depreciation	2,554	2,333
	$ 83,184	$ 140,325
	$ 83,184	$16,197,032
PREPAID EXPENSES	$ 345	$ 383
	$19,710,967	$19,999,719

LIABILITIES:	1963	1962
CURRENT LIABILITIES:		
Accounts payable	$ 14,711	$ —
Accrued federal and state income taxes (Note 2)	365,980	—
Other taxes accrued	44,510	—
Total current liabilities	$ 425,201	—
Deferred rent income, Lehigh and Susquehanna Railroad system	$ 190,290	$ 190,210
CAPITAL:		
CAPITAL STOCK, $1 par value:		
Authorized 3,000,000 shares		
Outstanding 2,202,127 shares	2,202,127	2,202,127
CAPITAL SURPLUS (Note 4), as annexed	17,462,956	17,607,382
EARNINGS RETAINED in the business, as annexed	—	—
	19,665,083	19,809,509
Less capital stock held in treasury, 78,790 shares, at cost	569,607	—
	$19,095,476	$19,809,509
	$19,710,967	$19,999,719

The accompanying notes are an integral part of the financial statements.

The accompanying notes are an integral part of the financial statements.

Table V. The Lehigh Coal and Navigation Company and Subsidiary Companies Consolidated Balance Sheets, May 31, 1962.

the Company which would expire at the end of the 35-year lease of the Lehigh and Susquehanna Railroad properties to the Reading Company. In Thompson's 1963 annual report, he made the following significant statement: "After a thorough exploration, the Board of Managers has come to the conclusion that efforts in this direction should be discontinued because a policy of paying out the greatest portion possible of the cash generated, within the limit of prudence, appears to be more advantageous from the stockholder's point of view."

The statement also referred to a normal yearly cash generation of 80 cents per share, arising from a lease rental of $2,283,482 per year which should permit the distribution of a yearly dividend of 60 cents per share, the major part of which should be free of normal Federal income tax.[179] For the first time, Model's policy of not reinvesting the funds generated from sale of assets and rental income was out in the open and only needed Board of Managers' approval to become company policy.

The Lehigh Coal and Navigation Company's management devoted most of its attention to the problems arising from the financial precariousness of the Central Railroad of New Jersey between 1964 and 1968:

1. In 1964, Thompson agreed to advance $500,000 at four percent interest to the Central Railroad of New Jersey to help finance coordination of facilities between the Lehigh Valley Railroad and the Central Railroad. The plan of coordination involved the common usage of certain Lehigh Valley Railroad trackage and certain parts of the Lehigh and Susquehanna Railroad property in order to effect substantial savings to the Central Railroad.[180]

2. On December 10, 1964, a decision was made to defer rent in the amount of $433,000 for each of the calendar years 1965 and 1966. It was also agreed that if the Central purchased the remaining leasehold interest before December 31, 1966, it would be relieved of paying the deferred amount.[181]

3. Negotiations with the Central toward its purchase of the Lehigh Coal and Navigation Company's reserved leasehold interest in the Lehigh and Susquehanna Railroad property proved fruitless and were discontinued in 1966.[182]

4. Quarterly rental due from the Central on April 1, 1966, was not paid because of insufficient funds. A plan of financial aid was subsequently worked out, and delinquent rental was paid July 1, 1966.[183]

5. The Central Railroad of New Jersey filed a petition for reorganization in Federal Court on March 22, 1967. As a consequence of this action, the Central did not make the rental payment due April 1, 1967, nor any to the Lehigh Coal and Navigation Company during 1968 and 1969. Also, a $500,000 note of the Central was written off as worthless.[184]

6. A favorable ruling by the Internal Revenue Service established the value of the Lehigh and Susquehanna leasehold interest for tax purposes to

be $38,064,494, with a book value of $17,335,678. This asset was to be amortized over a 35-year period from May 9, 1963, for both book and tax purposes. Amortization of this asset, amounting to $495,305 annually, provided the tax-free nature of a good portion of the dividends paid to the stockholders. Of the 60 cent dividend paid in 1966, 52.8 cents were tax free.[185]

Because rental payments from the Central were reduced in 1967 and then completely stopped in subsequent years, net profits of the Lehigh Coal and Navigation Company declined from $1,593,716 in 1964 to a loss of $854,942 in 1969. The 1967 annual report stated, "No action was taken on a dividend in May 1967 in view of the present status of the Central Railroad Company of New Jersey."[186] Dividend payments ceased to be made from this point on.

With the loss of rentals for the Lehigh and Susquehanna Railroad properties (the main source of income), the management of the Lehigh Coal and Navigation Company made a dramatic policy change in 1969. A decision was made to revitalize the Company and make it once again a major, profitable, and growing organization. Peter H. Engel, the new President of the Company, told the stockholders that "we have determined to concentrate the attention of management on the confectionery industry which we believe to offer substantial potential for profitable growth." Thomas R. Anderson of Minneapolis, Chairman of the Board, was largely responsible for encouraging this new direction in 1969.[187]

Coincidental to the entrance of the Company into the confectionery industry, Louis H. Thompson retired as president. At the same time, the remaining coal and transportation oriented members of the Board of Managers, G. Peter Fleck, Charles E. Oakes, and Carroll R. Wetzel, resigned from the Board.[188]

So ended the "railroad and coal" era of the Lehigh Coal and Navigation Company. The annual report to stockholders for 1970 had the caption, "The Lehigh Coal and Navigation Company and its subsidiary, Candy Corporation of America." The once famous "Old Company's Lehigh" was now headed in a completely new direction with a new management team. Coal had given way to candy.

L.N.C. Corporation

When the Plan of Reorganization of the Lehigh Coal and Navigation Company was carried out on May 31, 1962, assets with a net book value of $22,781,071 were transferred to the L.N.C. Corporation in return for 100 percent of the stock of this new Company. The stock of the L.N.C. Corporation was then distributed to the shareholders of the Lehigh Coal and Navigation Company.[189]

Assets in the L.N.C. Corporation consisted of the anthracite coal land and mining property, bituminous coal lands, railroad property and equipment, water utility property, Split Rock Lodge and adjacent lands in the Lake Harmony region, and other property spread over a wide area of the Pocono Mountains.[190]

As soon as the Company was formed, liquidation of the various assets was started. Sale of the diesel locomotives was completed by May 31, 1963. The 50-ton coal cars and remaining assets of the Lehigh and New England Railroad Company were sold in 1964, as was the Panther Valley Water Company's domestic water system. On January 29, 1965, sale of the bituminous coal properties owned by the Lehigh-Boone Land Company, a subsidiary, was completed.[191]

Finally, on May 16, 1966, the anthracite properties, leased under a long-term lease to the Greenwood Corporation, were sold to that company for $1,476,723.[192]

Pursuant to the plan of complete liquidation adopted by the shareholders on September 28, 1965, the L.N.C. Corporation made the following distributions: $5.00 per share on August 17, 1964, in partial liquidation; $2.50 per share on October 25, 1965, under the plan of complete liquidation; and on September 23, 1966, $1.00 per share together with the shares of Split Rock Lodge, Inc., and Blue Ridge Real Estate Company. The distribution of the stock of Split Rock Lodge, Inc., and Blue Ridge Real Estate Company took the form of a "unit certificate" consisting of shares of both companies. The unit certificates were traded in the Over-the-Counter market.[193]

There were no further annual shareholders' meetings of the L.N.C. Corporation. The directors, acting as liquidators, reported to the shareholders from time to time as any significant development occurred.[194]

Thus the Lehigh Coal and Navigation Company stockholders as of May 31, 1962, received one share of L.N.C. Corporation stock for each share of Lehigh Coal and Navigation Company stock. By September 23, 1966, each L.N.C. Corporation shareholder received $8.50 liquidating dividend for each share of that stock, plus one share of stock in Split Rock Lodge, Inc., and Blue Ridge Real Estate Company. In 1983 Blue Ridge/Big Boulder Corporation stock traded over the counter at a high asked price of 8½ and a low bid price of 7 per share.[195]

EPILOGUE

What finally happened to the leasehold interest in the Lehigh and Susquehanna Railroad System retained by the Lehigh Coal and Navigation Company? Annual rentals of $2,292,157 were to be paid for 35 years from May 9, 1963, under this arrangement. When the Plan of Reorganization was adopted by the Lehigh Coal and Navigation Company in May, 1962, the principal remaining asset in this Company was its remaining leasehold interest in the Lehigh and Susquehanna Railroad System. This rental constituted the main source of income for the company and was the cornerstone of its financial planning strategy, which had to be changed when the Central Railroad of New Jersey went into bankruptcy and failed to pay rentals beginning April 1, 1967.[196]

The next development on the Lehigh and Susquehanna leasehold matter occurred in September, 1972, when the Company entered into a lease and option agreement with the Lehigh Valley Railroad Company. Under this lease, the company received $585,000 rental in the first year and $575,000 in each subsequent year until 1998.[197] In addition, the Lehigh Valley Railroad was given an option to purchase the Company's interest in the Lehigh and Susquehanna Railroad properties within the first five years of the lease for a price of $7,500,000, less $125,000 of the annual rental paid in each year prior to the purchase. A settlement was also worked out with the Trustee of Central Railroad of New Jersey to pay $500,000 for all claims that either party may have against the other arising out of the Lehigh and Susquehanna Railroad lease or the operation of the Lehigh and Susquehanna Railroad by the Central or its Trustee.[198]

By May 31, 1972, the Lehigh Coal and Navigation Company had consolidated net operating loss carry forwards of approximately $12,608,000 available to reduce future consolidated taxable income.[199] Undoubtedly, this potential reduction in Federal taxes was one of the major reasons for the new direction of the Company in 1969 when it became obvious that rentals from the Lehigh and Susquehanna Railroad properties could no longer be depended upon as the major source of income. The confectionery industry was selected at that time to be this new source of income and thus make it possible to utilize the available loss carry forwards to reduce Federal income taxes.

However, the profits from the Candy Corporation of America were not as sweet as expected, and the Company was concerned about being unable to use the loss carry forwards which would completely expire by 1977.[200] Once again plans for rejuvenating the Company failed to materialize.

The new direction of the Company, its acquisition of several candy companies, led to operating losses until 1974, when a small profit of $57,347 was realized.[201] The small profit figures fell far short of the revenues expected when the Company made the dramatic policy change by concentrating management's attention on the candy industry. The company was fortunate to receive the $575,000 annual rental for the leasehold interest in the Lehigh and Susquehanna Railroad properties from 1972 to June 29, 1978, when the company's leasehold interest in this railroad property was finally sold for $5,250,000 to the Consolidated Rail Corporation.[202] The company had also received $120,000 as its share of the proceeds of sale of salvage on the abandoned segment of the Lehigh and Susquehanna property. With the proceeds received from the sale, the company retired its total consolidated debt and paid accumulated dividends on its Cumulative Convertible Preference Stock, Series B.[203]

However, the annual rental of the Lehigh and Susquehanna Railroad properties was no longer available to sustain the Company during hard times. The Company now had to exist from income arising solely from the candy business.

What finally happened to the anthracite coal properties that Josiah White demonstrated such ingenuity in developing in the early 1800's? These coal properties had been transferred along with other assets to the L.N.C. Corporation when the Plan of Reorganization was adopted on May 31, 1962, and were eventually sold on May 16, 1966, to the Greenwood Corporation, the company which had been leasing the properties, for $1,476,723.[204] The Greenwood Corporation was owned by the Fauzio brothers of Nesquehoning, Pennsylvania.

The Joseph Fauzio family of New Columbus, Nesquehoning, Pennsylvania, consisted of five sons and three daughters. The father and the sons had worked at the Lehigh Navigation Coal Company operation as hourly employees until the early 1930's, when the family purchased stripping and hauling equipment from two failing contractors, William Rousch and Joseph Petrosky. The fortunes of the family then steadily improved, as Fauzio Stripping and Hauling Contractors prospered.[205]

The brothers divided the responsibilities of running the family business. James headed the company. Frank (Patty) was in charge of operations. Bill, the oldest, supervised the shop where the stripping and hauling equipment was repaired. Patsy and Paul carried out various duties. Rose Fauzio, their mother, was a dominant influence in the business, but Joseph, their father, never took an active interest. The family business continued to grow and it was the largest stripping contractor working for Lehigh Navigation Coal Company when it ceased mining operations in 1954.[206]

Because the Fauzio Brothers had invested millions of dollars in stripping and hauling equipment over the years, it had a large stake in the welfare of mining operations in the Panther Valley. Accordingly, after the Lehigh Navigation Coal Company had been closed for approximately six

Fauzio Brothers stripping operation and equipment.

months in 1954, James and Frank were happy to join with Parton and Crane to organize the Panther Valley Coal Company to operate the Lansford Colliery (including the Nesquehoning Mines and Strippings) under lease from the Lehigh Coal and Navigation Company. When the Panther Valley Coal Company was acquired by the Coaldale Mining Company, the Fauzio Brothers continued to perform contract stripping work for these companies until each company terminated its lease. The Coaldale Mining Company was the last lessee. It ceased to operate the mining property at Coaldale on February, 1960.[207] At that time, the Lehigh Coal and Navigation Company entered into an agreement with the Fauzio Brothers, then operating under the name of Greenwood Stripping Corporation, to lease all of the Company's anthracite properties. The new lessee constructed a modern coal preparation plant, the Greenwood Breaker, on the site of the old Tamaqua Breaker. This lease was designed ultimately to realize the net carrying value of the anthracite properties, $4,000,000, to Lehigh Coal and Navigation Company.[208]

The Greenwood Stripping Corporation continued to lease the anthracite properties until May 16, 1966, when it bought the properties for $1,476,723.[209]

The Fauzio family then owned the coal lands and continued to produce coal from the various strippings, principally the Forty Foot and Mammoth Stripping, until 1974. At that time, Fauzio interests sold the entire anthracite property to the Bethlehem Mining Corporation, a subsidiary of Bethlehem Steel Company, Bethlehem, Pennsylvania, for a reputed $20,000,000.[210]

The Fauzio family thus played a significant role in the more recent history of the coal properties of the once famous Old Company's Lehigh.

The Bethlehem Mines Corporation is currently (1986) producing coal from the property and has benefited by the resurgence in demand for low sulfur coal for both space heating and metallurgical use. Because of the serious energy problems facing the United States, as well as other countries, arising out of our dependence for much of our oil supplies from the unstable Middle East, the coal industry, both anthracite and bituminous, has been rejuvenated. Demand for export sales is heavy and U.S. ports are unable to handle adequately the large tonnages which foreign countries want to purchase. Local sales have increased dramatically as more public utilities and industries revert to coal. Thus, R.V. White's abiding faith that anthracite coal would once again come into good times might very well be borne out even though it took a period of three decades for it to happen.

It is a shame that the Lehigh Navigation Coal Company wasn't able to alter its operations in 1954 so that it could survive and ultimately prosper until the present time. Where does the blame lie? Was it the fault of the United Mine Workers of America and the Company's employees who didn't want the Company to reduce the size of its operations, even though the Company would have been better suited to the shrinking market for anthracite? Was it the Company's submissive management which had given in to the Union too often and was unable to provide the necessary leadership? Or, was it the fault of certain stockholders who selfishly decided that their stock holdings could be worth more by liquidating this asset-rich company? Possibly the answer is that everyone, employees, management, and stockholders alike, was jointly responsible for the demise of the venerable Lehigh Coal and Navigation Company.

When this book was about to go to press in 1986, the final curtain was dropped on the Lehigh Coal and Navigation Company. As referred to previously, this famous old industrial concern had been reduced to its last subsidiary, Cella's Confection, Inc., a candy-making firm that specialized in chocolate-covered cherries. The shareholders of the company recently received a proxy statement informing them that the annual meeting of the shareholders would be held in the Union League Club in New York on July 29, 1985. The primary purpose of the meeting was stated "to discuss the advisability of voluntarily dissolving the company and take action upon a resolution to adopt a Plan of Complete Liquidation and Dissolution of the

Lehigh Coal and Navigation Company." The assets of its last subsidiary, Cella's Confection, Inc., were sold to Tootsie Rolls.[211]

The Lehigh Coal and Navigation Company no longer exists. It should be remembered for its great contribution to the rise of our youthful nation from 1820 to 1969 as a famous coal and transportation entity.

NOTES

1. Thomas C. James, "A Brief Account of the Discovery of Anthracite Coal on the Lehigh," *Historical Society of Pennsylvania Memoirs*, Vol. 1, Pt. 2 (Philadelphia: Carey, Lea, and Carey, 1826), 318. Dr. James used the family name as Ginter, but the monument erected at Summit Hill, commemorating the 150th anniversary (1941) of the discovery of anthracite, was inscribed "Ginder." For more information on Philip Ginter (Ginder) see Geo. Korson, *Black Rock* (Baltimore: John Hopkins University Press, 1960), 1-31.

2. Ibid., 319.

3. History of the Lehigh Coal and Navigation Company (Philadelphia: W.S. Young, 1840), 3, hereafter referred to as *Lehigh History*.

4. Richard Richardson, *Memoir of Josiah White* (Philadelphia: J.B. Lippincott and Company, 1873), 40.

5. Lehigh History, 5.

6. *Pennsylvania Legislative Acts, 1818* (Harrisburg: C. Gleim, 1818), 205.

7. *Lehigh History*, 9.

8. The Story of The Old Company: Written and Published as a Contribution to the Celebration of the One Hundred and Fiftieth Anniversary of the Discovery of Coal at Summit Hill, August 28 to September 1, 1941 (Easton, PA: Mack Printing Company, 1941), 14, hereafter referred to as *The Story of The Old Company*.

9. Lehigh History, 14.

10. The Story of The Old Company, 18.

11. *Lehigh Coal and Navigation Company, Annual Report* for 1829 (Philadelphia: T.A. Conrad, 1830), 12, hereafter referred to as *L.C.& N. Co. Annual Report*.

12. A.E. Wagner, *History, Government and Geography of Carbon County, Pennsylvania* (Allentown: Press of Berkemeyer, Keck & Co., 1910), 44, hereafter referred to as *History of Carbon County*.

13. Ibid.

14. Ibid.

15. Ibid., 45.

16. Ibid., 46.

17. Ibid.

18. J.B. Warriner, "The Lehigh Navigation Coal Company," *The Mining Congress Journal*, July 1930, 579.

19. Ibid.

20. Ibid.

21. The Story of The Old Company, 15.

22. Ibid., 54, 56.

23. The "Old Company," *Fifteenth Annual Model Mining Issue, Coal Age*, Dec. 1939; (New York), 486.

24. L.C.& N. Co. Annual Report for 1939 (Philadelphia), 24.

25. Ibid., 7, 8.

26. The Story of The Old Company, 15.

27. Warriner, "Coal Company," 580.
28. Interviews by the author with H. Louis Thompson, retired president of the Lehigh Coal and Navigation Company. Interviews took place at various times in 1981 and 1982 at Bethlehem, Pa. Hereafter referred to as Thompson Interviews.
29. The Story of The Old Company, 31.
30. Ibid., 47.
31. Ibid.
32. Ibid., 48.
33. "Who's Who - in Coal Mining," *Coal Age*, May 1928, 35.
34. "J.B. Warriner, Vice-President and General Manager, The Lehigh Coal and Navigation Company: A Biography," *The Explosives Engineer*, May 1929 (Wilmington, Delaware), 161.
35. The Story of The Old Company, 49.
36. *L.C.& N. Co. Annual Report for 1935* (Philadelphia), 22.
37. Ibid.
38. *L.C.& N. Co. Annual Report for 1936* (Philadelphia), 22.
39. *L.C.& N. Co. Annual Report for 1937* (Philadelphia), 11.
40. *Who's Who in America, Volume 29*, Marquis -Who's Who (Chicago, 1956-1957), 2760.
41. According to personal knowledge of the author, substantiated by L.C.& N. Co. sales and records in possession of T.R. Berger, Allentown, Pa. Hereafter referred to as author recollections.
42. Ibid.
43. *L.C.& N. Co. Annual Report for 1938* (Philadelphia), 15.
44. *L.C.& N. Co. Annual Report for 1939* (Philadelphia), 5.
45. *L.C.& N. Co. Annual Report for 1940* (Philadelphia), 28.
46. Ibid.
47. *L.C.& N. Co. Annual Report for 1941* (Philadelphia), 13.
48. Ibid., 5.
49. Ibid., 8, 9, 10.
50. Ibid., 13.
51. Ibid., 9.
52. Ibid., 7.
53. *L.C.& N. Co. Annual Report for 1943* (Philadelphia), 14.
55. *L.C.& N. Co. Annual Report for 1944* (Philadelphia), 6.
56. Ibid., 6, 7.
57. Ibid., 15.
58. *L.C.& N. Co. Annual Report for 1945* (Philadelphia), 4.
59. Ibid., 7.
60. Ibid., 15.
61. Berger files.
62. Ibid.
63. Undated newspaper article from personal files of Emily Evans (Pottsville, Pa.).

64. Data summarized from *L.C.& N. Co. Annual Reports from 1946 through 1954* (Philadelphia).

65. *L.C.& N. Co. Annual Report for 1949* (Philadelphia), 12.

66. Berger Files.

67. *L.C.& N. Co. Annual Report for 1949,* op. cit., 6, 7, 8, 9.

68. Ibid., 6.

69. Ibid., 8.

70. From knowledge gained in conversations with executives of other major coal companies during this period.

71. *Foreman's Guide on Consideration Mining,* Lehigh Navigation Coal Company, 7. (In author's personal collection.)

72. Ibid., 8, 9.

73. Ibid., 10.

74. Berger files.

75. Author was one of the Company's representatives at the U.M.W.A. Washington meeting.

76. *Agreement between Lehigh Navigation Coal Company and United Mine Workers of America,* rate sheets and specifications, 1952.

77. *L.C.& N. Co. Annual Report for 1949,* 12, 13.

78. Ibid., 5, 6.

79. *L.C.& N. Co. Annual Report for 1950,* 5.

80. Berger Files - news release.

81. Berger Files.

82. *L.C.& N. Co. Annual Report for 1951,* 5.

83. Thompson Interviews.

84. *L.C.& N. Co. Annual Report for 1951,* 6.

85. *Lehigh Navigation Coal Company Financial Statements* for 1951 and 1952, 6.

86. Berger files.

87. Thompson Interviews.

88. *L.C.& N. Co. Annual Report for 1952,* 1.

89. Ibid.

90. Ibid., 11, 14.

91. Letter to stockholders, Oct. 23, 1952.

92. *L.C.& N. Co. Annual Report for 1952,* 14.

93. *L.C.& N. Co. Annual Report for 1953,* 3.

94. Ibid., 10.

95. Berger files.

96. Ibid.

97. *L.C.& N. Co. Financial Statements* for 1952-1953, 6.

98. *L.C.& N. Co. Annual Report for 1953,* 3, 4.

99. Ibid., 11.

100. Ibid., 10.

101. Thompson Interviews.

102. *Solid Fuel Retailer*, Feb. 1954.

103. Berger files.

104. Berger files, Panther Valley Mine Closing, 1954.

105. Berger files.

106. J.J. Crane, *Chronological Outline of Events in Panther Valley* (unpublished internal company memorandum in possession of the author), 3.

107. Ibid.

108. Ibid.

109. Ibid., 4.

110. Ibid.

111. Ibid.

112. Ibid.

113. Ibid.

114. Ibid., 5.

115. Ibid.

116. Ibid.

117. Ibid.

118. Ibid.

119. Ibid., 6.

120. Ibid.

121. Author recollections.

122. Agreement between Panther Valley Coal Co., Inc., and United Mine Workers of America (Locals Nos. 1708 and 1738) Rate Sheets and Specifications, September 7, 1954 (Lansford, Pa.).

123. Berger files.

124. Information provided to C. Millard Dodson and H. Louis Thompson by James H. Pierce and Larry Rouselle, officials of the Coaldale Mining Company. When Coaldale Mining Company negotiated to lease the Coaldale Colliery, the lessor requested to know the source of financing. This information was personally relayed to the author.

125. *L.C.& N. Co. Annual Report for 1954*, 3.

126. Ibid., 6-14.

127. Ibid., 10.

128. Ibid., 8, 9

129. *L.C.& N. Co. Financial Statements* for 1954.

130. *L.C.& N. Co. Annual Report for 1954*, 6.

131. *L.C.& N. Co. Annual Report for 1955*, 2 and 6.

132. Author recollections.

133. Ibid.

134. Ibid.

135. Copy of Agreement of Sale of Common Stock of the Panther Valley Coal Company to Coaldale Mining Company, dated May 1955 (in possession of author).

136. *L.C.& N. Co. Annual Report for 1955*, 3.

137. Ibid., 4.
138. Ibid., 5.
139. Ibid., 6.
140. Ibid., 2.
141. Ibid., 6.
142. *L.C.& N. Co. Annual Report for 1957*, 12, 13.
143. Ibid., 11.
144. Ibid., 2.
145. Ibid., 6-7.
146. Ibid., 7.
147. Ibid.
148. Ibid., 10.
149. Ibid., 20.
150. Ibid., 2.
151. *L.C.& N. Co. Annual Report for 1958*, 2.
152. Ibid., 3.
153. Ibid.
154. Thompson Interviews.
155. Randolph Kulp, *The Lehigh and New England R.R.*, Lehigh Valley Chapter N.R.H.S., 1967, 20.
156. *L.C.& N. Co. Annual Report for 1959*, 2.
157. Ibid.
158. Ibid.
159. *L.C.& N. Co. Annual Report for 1960*, 3.
160. Ibid.
161. Ibid., 2.
162. Ibid., 3.
163. Ibid.
164. Thompson Interviews.
165. *L.C.& N. Co. Annual Report for 1961*, 1.
166. Ibid.
167. Ibid.
168. Ibid., 2.
169. Ibid.
170. Ibid., 3.
171. Ibid.
172. Ibid., 2.
173. Ibid., 3.
174. Ibid., 6.
175. Ibid., 7.
176. Ibid., 2.

177. *The Lehigh Coal and Navigation Company Annual Report for 1963* (Bethlehem, Pa.), 1, hereafter referred to as the New L.C.& N. Co.

178. *The New L.C.& N. Co. Annual Report for 1967*, Fly leaf.

179. *The New L.C.& N. Co. Annual Report for 1963*, 1.

180. *The New L.C.& N. Co. Annual Report for 1964*, 1.

181. *The New L.C.& N. Co. Annual Report for 1965*, 1.

182. *The New L.C.& N. Co. Annual Report for 1966*, 1.

183. Ibid.

184. *The New L.C.& N. Co. Annual Report for 1967*, Fly leaf.

185. Ibid.

186. Ibid.

187. *The New L.C.& N. Co. Annual Report for 1968*, 1, 2.

188. Ibid., 2.

189. *The New L.C.& N. Co. Annual Report for 1969*, 1, 2.

190. *L.N.C. Corp. Annual Report for 1963*, 1.

191. *L.N.C. Corp. Annual Report for 1965*, Fly leaf.

192. *L.N.C. Corp. Annual Report for 1966*, Fly leaf.

193. Ibid.

194. Ibid.

195. Blue Ridge Real Estate Company Big Boulder Corporation Annual Report for 1983 (Blakeslee, Pa.), 13.

196. *L.C.& N. Co. Annual Report for 1967*, Fly leaf.

197. *The New L.C.& N. Co. Annual Report for for 1972*, 1.

198. *The New L.C.& N. Co. Annual Report for 1971*, 2.

199. *The New L.C.& N. Co. Annual Report for 1972*, op. cit., 16.

200. Ibid.

201. *The New L.C.& N. Co. Annual Report for 1974*, 1.

202. *The New L.C.& N. Co. Annual Report for 1978*, 1.

203. Ibid.

204. *L.N.C. Corp. Annual Report for 1966*, Fly leaf.

205. Author's personal knowledge of Fauzio family supported by interview with George McDonald, Lansford, Pa., in 1983.

206. Author's recollections.

207. *L.N.C. Corp. Annual Report for 1960*, 3.

208. Ibid.

209. *L.N.C. Corp. Annual Report for 1966*, Fly leaf.

210. Thompson Interviews.

211. Lehigh Navigation Company Proxy Statement for the Annual Meeting of Shareholders to be held July 29, 1985, at the Union League Club, New York, New York.

INDEX

A

accident rate 57
advertising campaigns 35, 41
aggregate plant 48–50, 62
aggregate, development of 48
Anthra-Tube 41
anthracite. *See also* Mining, Transportation
  demand for 47, 55, 110
    during war 39–41, 44, 59
    post-war 61
    decline in 28, 55, 62, 63, 65, 68
  field 19
  production 26, 28–30, 35, 37, 39–42,
    55, 57–58, 61, 62, 65–66, 84
    post-war 44, 47, 59–60
  promotion of 35, 41, 50, 61
  sales of 35, 47, 62, 63, 64, 82, 83
    affecting equalization 44
Anthracite Conciliation Board 27, 28, 60
Anthracite Institute 41
Ashley Planes 9
assets 19–21, 35, 63, 68, 84, 91, 100–1
  following liquidations 100
  income from sale of 103
  liquidation of 96, 99, 105
  railroad leases 63
  real estate 62, 90, 97, 100

B

balance sheets. *See* tables II, IV, V
Banta, Walter 62
Beckwith, Albert 33
beneficiation process 42, 89
Berger, T.R. 24, 35
Bethlehem Mining Corporation 110
Bethlehem Steel Co., 110
Big Boulder 97
bituminous mine, purchase of 91
Blue Ridge Real Estate Co. 100, 105
Board of Conciliation. *See* Conciliation
  Board
Board of Managers
  changes on 64–65, 67–68, 96
  conflict on 93–97
Bolinger, John C. 34, 65, 71
Boone County Coal Company 93, 94
  sale of 96
Boone County Coal Corporation 91, 92,
  96

bootleg mines 28–29, 65
Boyd, Robert L. 34
Boyles, Vernon 34
Brennan, Martin 65, 72, 75, 76, 78, 79
Bright's Department Store 25
Broad Mountain 19
Bundy Company 25

C

canal transportation of coal 5, 9. *See
  also* Lehigh Canal
Candy Corporation of America 102,
  105
cement hauling by Lehigh and New England RR 39, 50, 63, 68, 89, 94
cement plants 19, 89, 94
Central RR of Pennsylvania 42, 91, 93,
  94
Central RR of New Jersey 6, 8, 9, 19,
  21, 37, 42, 63, 68, 69, 89, 92, 103
  bankruptcy of 38, 107
check off system 27, 30
Clark, E.W. 13
closings. *See* Mine closings; Consolidation
coal. *See also* Anthracite
  transportation of. *See* Transportation
  use of for industrialization 1, 94, 110
coal measures, diagram of 14
coal mine company 1. *See also* Lehigh
  Coal Mine Company
coal mines 19
Coaldale 25, 66
  breaker 74
  colliery 16, 37, 42, 50, 83
  Local 77
Coaldale Mining Company 83, 86, 88,
  92, 94, 109
Coaldale Observer 25, 30
collieries. *See* Coaldale, Greenwood,
  Lansford, Nesquehoning, Tamaqua
colliery closings 72. *See also* Mining operations
Community Park 22
"company man" 24
company towns 25
competition
  from other coal producers 86
  from other fuels 45, 63, 65, 86
Conciliation Board 27, 28, 60

concrete made with aggregate 48. *See also* Cement

confectionery industry 104, 107, 108, 110

conflict
of interests 21, 60
on Board of Managers 93, 95, 96

consideration work 58–59, 72. *See also* Pay rates

consolidation as cost-cutting measure 37, 55, 63, 65–66, 75

consolidation plan 76. *See also* Reorganization

contract pay rates 58–60, 75

cost reduction measures 63–65, 75–76, 110

Cox, James 13

Cox, John 13

Crane, Joseph J. 71, 73, 81, 82, 109

D

debt 35, 42, 44, 51, 54, 61, 84, 100

Dechert, Robert 67

decline. *See also* Anthracite
in demand 47, 55, 63, 65, 68, 83, 88, 93, 94.
in productivity 60–61

development of real estate 62

diversification 91, 93, 95

dividends 38–41, 47, 51, 54, 61–64, 67, 69–95, 85, 86, 90, 93, 95, 104, 108. *See also* Stocks

Dodson breaker 66

Dodson, C. Millard 64, 67, 71, 76, 80, 84–86, 90, 91, 93, 95, 96

Dodson years 71–97

Douglas, Edwin A. 13

Dumaine, Frederick C., Jr. 65, 96

E

Edison Anthracite Coal Co. 37, 44

Edwards, W.H. 29

employee relations 35

employment in coal industry 65–66

Engel, Peter H. 104

equalization 29–30, 44, 88

ethnic groups 55, 57

Evans, Evan 22, 23, 25, 37, 45, 47, 59–60, 62, 65, 66

F

Fauzio family 95, 108–110

Fauzio, James and Frank 81, 82

Financial statements 84. *See also* Balance sheets

Fine, Governor John S. 80

fires 42

Fleck, G. Peter 104

floods 9, 41–42

froth flotation plant 16, 42, 43, 50

G

Gates, John S. 34

General Crushed Stone Company 89

General Mine Committee 75–80

Gildea, James H. 25, 30, 80

Ginter, Philip 1

Glen Alden 55

government takeover of mines 40

gravity railroad 5

Greenwood breaker 109

Greenwood colliery 37, 39

Greenwood Corporation 105, 108

Greenwood stripping 44, 47

Greenwood Stripping Corp. 95, 100, 109

growth in coal production 13

H

Hanks, J.S. 34

Harris, J.S. 13

Hauto yard 18

Hauto power plant 20

Hauto, George F. 2

Hauto Tunnel 6

Hazard, Erskine 1, 9, 13

Helms, D.C. 63

Hickey, William M. 64, 96

holding company 19

Hudson Coal Company 55

I

illnesses, occupational 57

immigrants 21, 23

improvements 50, 55, 64

improvements at mines 41, 63, 68, 75

industrial market for anthracite 61, 64

J

job losses 65–66

K

Kennedy, Thomas 74, 75, 76, 77, 82

Kidd, Glenn O. 34, 47, 62, 65–66, 67
resignation of 71

Kuebler, Charles S. 33

L

L.N.C. Corporation 104, 108
  transfer of assets to 101
labor. *See also* Union
  contract during war 40, 45
  problems 47, 55, 60, 77, 79, 80, 88
  unrest 28, 29–30, 35
Lake Harmony 40
Lansford 5, 19, 25, 41
  colliery 17, 50, 55, 81-82, 109
  mines 65, 66, 74, 88
Lansford shops 17
  consolidation at 65
Lathrop, W.A. 13, 27
layoffs 74
Lehigh and New England Railroad 9, 19,
  37, 38, 39, 41, 50, 60, 61, 62, 69, 89,
  92, 93, 94, 96
  as cement hauler 39, 69, 92, 94
  dieselization of 49, 62
  sale of 94, 96, 99
Lehigh and Susquehanna Railroad 9, 19,
  21, 37, 38, 42, 68, 89, 91, 92, 93,
  100, 101, 107
  sale of 63–64, 96, 101, 108
  tax value of 63, 104
Lehigh Canal 37
Lehigh Coal and Navigation Company
  2, **13–31**, 27, 31, 41, 42
  as candy manufacturer 104, 107–108,
  110–111
  in 1939 19
  1962 assets of 100–101
  liquidation of 110
  losses by 61
  operators of railroad 9
  presidents of 13, 31, 33, 34, 104
  reorganization of 71, 100–101,
  104–105, 107. *See also* "Plan 2"
  securities, 1939 (table) 15
Lehigh Coal Company 2
Lehigh Coal Mine Company 1, 2
  losses by 65–66
Lehigh Navigation Coal Company 19,
  25–26, 28, 42, 44–45, 50, 110
  as employer 25
  changes in personnel at 71
  during White's presidency 33
  losses by 35, 37, 61, 63, 66, 68, 85
  during Depression 31
  mine closings. *See* Mine closings

mining engineers at 34
personnel changes at 63
presidents of 27, 34, 45, 65, 67–68,
  71
profits by 37, 38, 55, 62
  during war 41
strikes against 21, 24
Lehigh Navigation Coal Sales Company
  82, 88
Lehigh Navigation Company, founda-
  tion of 2
Lehigh Navigation-Dodson Company
  100
Lehigh Valley Railroad 27, 103, 107
"Lehigh Weatherman" 35
Leisenring, E.B. 13, 64
Lelite 48–50, 62, 83
Lewis, John L. 40, 60, 73, 74, 76–78, 83
  at Tamaqua Local meeting 78
  meeting with General Mine Committee 79
liquidation 105, 110
losses 28, 29, 35, 37, 55, 63, 66–67, 68,
  83, 85, 88, 90, 93, 94, 104, 108
  during Depression 31

M

Mammoth stripping 44, 47, 110
management, modern methods of 34
management takeover 68
market study 62–63
marketing. *See* Anthracite, promotion of
McColough, C. Peter 34
McCready, James C. 80, 81
McGrath, John 33
merger 64
mine closings 65–66, 80–81
  reasons for 30
mine properties 109
miners 24
  as bootleggers 28
  in Panther Valley 58–61
mines 19
  bootleg 88
  deep 52, 74
  improvements at 37, 39, 64
  lease of 81–83, 86, 88, 89, 92, 94,
  95, 100, 105, 109
  maintenance of after closing 81, 89
  reopening, plans for 72–76, 80–81
mining immigrants 21
mining operations, suspension of
  71–72, 80–81

mining properties 81, 83, 86, 90, 95, 105
  value of 90
mining techniques 51, 54, 57–58, 72, 74
Model, Leo 63–64, 67, 91, 93, 95, 97,
  99, 103
Model, Roland and Stone 63–64
modernization 50, 54, 62, 63. *See also*
  Improvements
  post-war 50, 61
  financing of 50–51, 54
Molly McGuires 57
Morgan, Marshall S. 68

N

Nesquehoning 5, 25
  breaker 18, 55
  Colliery 36, 37, 44, 72
  mine 66, 81
  Railroad 8, 9, 92
New York, New Haven and Hartford Rail-
  road 19
Newbold, Richard C. 65, 66
Nuelle, J.H. 13, 30

O

Oakes, Charles E. 104
O'Donnell, Ambrose 80
"Old Company's Lehigh" coal 100
Old Timers Club 22, 23

P

Panther Valley
  industrial development in 25, 74
  miners 55, 57–61
  mines 5, 20, 29, 35, 58, 72, 74, 100,
    109
    closing of 80–81. *See also* Mine clos-
      ings
    sketch of 56
  people of 21, 24, 80, 81
  politics in 25
Panther Valley Coal Company 81, 83,
  86, 88, 92, 109
  losses by 88
  merger with Coaldale 88
Panther Valley General Mine Committee
  75–80
Panther Valley Water Company 20
Pardee, Calvin 13
Parton, W. Julian 33, 34, 63, 65, 71, 73,
  81, 89, 109
pay rates 58–60, 75. *See also* Wages

Penn Navigation Company 91, 97, 100
picketing 77–81, 88
piece work 58. *See also* Pay rates
Pierce, James H. 83, 88
Pisgah, Mt. 5, 19
"Plan 2" 73–79
Planes, Ashley 9
Pocono resort area 20, 40, 97
pollution laws 42
PP&L 63, 87
Presidents of Company 13, 99
production. *See* Anthracite production.
productivity 44, 63, 72–75, 83, 85
  decline in 44–45, 59–61
  need for increased 72–75
profits 28, 37–38, 41, 42, 45, 47, 62,
    64, 89–90, 92, 95, 100, 104, 108
  from railroad 63
  from railroad leases 37
  during war 41

R

radio, use of in sales campaigns 35
Railroad, Summit Hill 5
railroad, Nesquehoning to Lansford
    53, 55
railroads 9, 19, 21
  leases of 21, 37, 38, 42, 63, 89, 91,
    100–101, 103–104, 107–108
  profits from 37–38, 63, 89, 94
  sale of 92
Reading Coal 55
Reading RR 101
real estate development 62
recession, effect of on LC&N 93
reclamation plant *See* Recovery plant
recovery equipment 50, 62
recovery plant 39, 42, 47–48
recreation 22, 23, 25, 97, 100–101,
    107. *See also* Sports
reorganization plans 71, 73–79,
    100–101, 104–105, 107
Richards, Norman 63
Riley, L.A. 13
Rutledge, John 34

S

safety programs 57
sales. *See also* Anthracite, sales of
  campaign 35, 62, 63, 64
  through Weston Dodson Co. 64
  decline in 47, 63, 65
Scattergood, Mr. 67

Schwable, Henry C.   34
securities   15
Sharp Mountain   1
shipping subsidiary   91, 97
Siefert, Harry W.   34
Smith, Bruce D.   64
Solid Fuels Administration   40
Split Rock Lodge   40, 62, 97, 100, 105
sports   23, 100. *See also* Recreation
steel strike, effect of on mines   94
Stevens, David (Red)   79
stock, Company   39, 63, 64, 68
   concentrations of   64, 68, 71
   delisted from Stock Exchange   101
   transactions, investigation of   78, 79
stockholders   38, 39, 61, 64, 68, 95, 110
   discord among   93, 95, 96
stocks *See* Assets; Dividends; Table I
strikes   24, 26, 28, 29, 35, 42, 47, 59–60, 75
   effect of on sales   60
   impact of on railroad revenues   21
stripping operations   43, 44, 47, 50, 54, 61, 74, 108–110
subsidiary companies   15, 19–20, 42, 64, 67, 91
   L.N.C. Corporation   101
   role of in conflict of interests   21
Summit Hill   2, 25
   gravity railroad   5
   Water Company   20
Switchback   5, 9, 10

T

Tamaqua   25
   breaker   74
   colliery   16, 37, 39, 41, 42, 49, 50, 89
   Local   77–78, 79, 80
   truck scale   53
Taney, Robert   48
tax liabilities   90, 107
Thayer, F.M.   34
Thomas, Evan   84
Thompson, H. Louis   21, 34, 62, 68, 93–96, **97–103**, 99, 104
   on R.V. White   68–69
Tout, Joseph (Buck)   81, 88
transportation of coal   2, 5, 9, 13, 21, 39, 41, 50
truck scale   53

U

union. *See also* General Mine Committee; United Mine Workers of America
   demands   27, 30
   grievances over pay   59–60
   vote on "Plan 2"   76–77
unionism   24, 81
unions, local   72, 77
   meet with Lewis   78
   meetings of   73, 75, 77, 80, 81
United Mine Workers of America   23, 24, 29–30, 40, 47, 60, 65–66, 71–72, 76, 82, 92, 110

V

Volpe, Santo   75

W

wages   29–30, 45, 60. *See also* Pay rates
Warriner, J.B.   23, 27, 28, 29, 33, 45
Warriner, S.D.   13, 26, 27, 30, 40
water companies   37
water supply   44
Weatherman, Lehigh   35
Weiss, Col. Jacob   1
Weissport   1
Weston Dodson Coal Company   64, 71
Wetzel, Carroll R.   104
White, Josiah   1, 2, 5, 9, 13
White, Robert V.   23, 29, 31, **33–69**, 35, 38–39, 41–42, 45, 47, 61, 62, 67–69, 97
   as consultant   67–68
   policies of questioned   64–65
   reasons for ouster of   67–69
   resignation of   67
Wilkes-Barre-Scranton Railroad   19, 37, 38
   sale of   91
work day   30
work week   40
workers' attitudes   55, 60, 72, 76, 78, 88
workforce, reduction in   65–66
working conditions, 19th century   57
World War II   39–42
   employees serving in   39, 42
   production during   39, 40, 41, 42, 59